FOREWORD

The Road Research Programme has two main fields of activity:

- promotion of international co-operation in the field of research on road transport, the co-ordination of research facilities available in Member countries and the scientific interpretation of the results of joint experiments;
- International Road Research Documentation, a co-operative scheme for the systematic exchange of information on scientific literature and of current research programmes in Member countries.

The present programme is primarily concerned with defining the scientific and technological basis needed to assist governments of Member countries in decision-making on the most urgent road problems:

- development and evaluation of integrated urban and suburban transport strategies, taking into account economic, social, energy and environmental requirements;
- traffic control systems, both in urban and rural areas, to optimise traffic operation and to enhance quality of service provided to road users;
- formulation, planning and implementation of common overall strategies for safety;
- planning, design and maintenance of the total road infrastructure, taking account of economic, social and technical developments and needs.

ABSTRACT

This study forms part of an ongoing co-operative research effort within the OECD Road Research Programme to advance the state-of-the-art of bridge management. As a follow-up to the 1976 report on Bridge Inspection and that of 1979 on the Evaluation of Load Carrying Capacity of Bridges, it addresses the issue of bridge maintenance, in particular the strategies to be considered, the most appropriate methods and techniques, the management policy to be defined and the major research needs in this field. Experts from sixteen Member countries participated and prepared this report on the basis of an international survey of present practice and ongoing research. The Group's recommendations call for an increased effort in this key sector of transport policy to maintain a high standard of serviceability of the existing highway network.

PREFACE

Since 1974 the OECD Road Research Programme has been involved in a series of studies on the management of existing road bridges and engineering structures.

The importance of the present study is self-evident. Bridges are crucial components in the road network; they are designed to meet traffic needs and represent a considerable asset both in economic terms and, frequently, in terms of their historical and cultural interest.

Two research groups have looked into the areas of "Inspection" and "Evaluation of Bearing Capacity" and it was thus a logical step to turn attention to the third element of bridge management, namely, maintenance, bearing in mind its evident links with the other components of the total management process.

Maintenance, which can be defined as embracing all the various operations designed to ensure appropriate serviceability, must meet the following requirements. It should:

- guarantee a satisfactory level of safety;
- ensure optimum traffic capacity;
- maintain the national stock of bridges in an optimal state.

These aims reflect the importance of what is at stake and hence the value of and need for a rational maintenance policy.

The report starts out by reviewing the present situation in the various countries, both in regard to maintenance methods and techniques and to the definition and implementation of a management policy.

The main conclusions of this review are that the resources available as regards policy are not sufficient and that there is a paucity of technical and economic data on the issues at hand; from the technological and methodological standpoint there is a need for improving existing techniques and developing new ones.

The report then goes on to discuss in some detail the various problems and aspects to be dealt with in the context of drawing up a maintenance policy. In the course of the study it emerged strikingly that there was a considerable degree of interaction between the various aspects of management and that problems tended to be global in nature. This led to the examination of certain general problems, particularly in their organisational and economic aspects. In its attempt to define a maintenance policy, the working group looked into the practical problems of devising a management policy and highlighted the complexity of the issues and the hurdles to be overcome.

In conclusion, the need to establish truly rational policies for bridge maintenance has recently become increasingly apparent both as a result of economic pressures and in response to the demands of users for quality of traffic service and safety. There is hence a need for international co-operation in a wide-ranging research effort, even if the problems posed are not necessarily the same from one country to another.

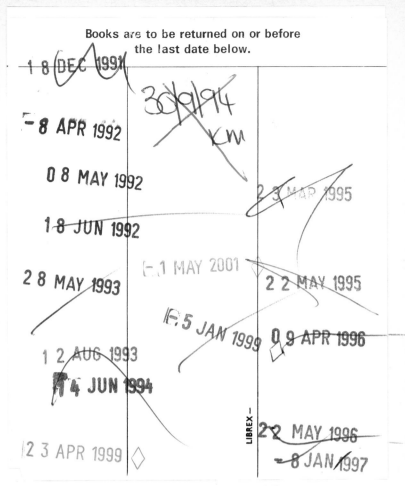

**A REPORT PREPARED
BY AN OECD ROAD RESEARCH GROUP**

SEPTEMBER 1981

ORGANISATION FOR ECONOMIC CO-OPERATION AND DEVELOPMENT

The Organisation for Economic Co-operation and Development (OECD) was set up under a Convention signed in Paris on 14th December 1960, which provides that the OECD shall promote policies designed:

— to achieve the highest sustainable economic growth and employment and a rising standard of living in Member countries, while maintaining financial stability, and thus to contribute to the development of the world economy;
— to contribute to sound economic expansion in Member as well as non-member countries in the process of economic development;
— to contribute to the expansion of world trade on a multilateral, non-discriminatory basis in accordance with international obligations.

The Members of OECD are Australia, Austria, Belgium, Canada, Denmark, Finland, France, the Federal Republic of Germany, Greece, Iceland, Ireland, Italy, Japan, Luxembourg, the Netherlands, New Zealand, Norway, Portugal, Spain, Sweden, Switzerland, Turkey, the United Kingdom and the United States.

Publié en français sous le titre :

ENTRETIEN DES OUVRAGES D'ART

* *
*

© OECD, 1981
Queries concerning permissions or translation rights should be addressed to:
Director of Information, OECD
2, rue André-Pascal, 75775 PARIS CEDEX 16, France.

TABLE OF CONTENTS

Chapter I

INTRODUCTION 7

I.1 Place of the Study in the Framework of Bridge Management 7
I.2 Terms of Reference 7
I.3 Statement of the Problem 8
I.4 General Maintenance Objectives 9
    I.4.1 Avoiding damage or injury to third parties 9
    I.4.2 Ensuring the best possible conditions for traffic 9
    I.4.3 Preserving the national bridge stock as effectively as possible 9
I.5 Some Aspects of Bridge Maintenance 10
    I.5.1 Overview of the problems 10
    I.5.2 Durability 10
    I.5.3 Preventive maintenance 10
    I.5.4 Problems posed by the economic approach 10
    I.5.5 Technical difficulties and uncertainties 11
I.6 Scope of the Study 11
    I.6.1 Definition of the field of bridge maintenance 11
    I.6.2 Nature and characteristics of the bridges concerned 12
I.7 Structure of the Report 13

Chapter II

MAINTENANCE POLICY - PRESENT SITUATION 14

II.1 General Concept of Maintenance Policy 14
    II.1.1 Aims of bridge maintenance 14
    II.1.2 Definitions of maintenance 16
    II.1.3 Preventive maintenance 16
    II.1.4 Some special aspects of maintenance 18
II.2 Categories of maintenance 18
II.3 Organisation of Bridge Maintenance 19
    II.3.1 Authorities responsible for bridge maintenance 19
    II.3.2 Personnel 19
    II.3.3 Preparation and approval of the maintenance programme 20
    II.3.4 Carrying out the maintenance work 21
II.4 Assessment of the Need for Maintenance 23
    II.4.1 Basis for assessment 23
    II.4.2 Engineering criteria 26
    II.4.3 Financial criteria 26

    II.5 The Replacement of Bridges 27
    II.6 The Cost of Maintenance 30
    Bibliography 33

Chapter III

    MAINTENANCE TECHNIQUES - CURRENT SITUATION 34
    III.1 Inventory of Maintenance Operations 34
    III.2 State of the Art of Certain Specialised Techniques 36
        III.2.1 Bridge surfacing and waterproofing 36
                Temporary interventions 39
                Definitive interventions 40
                Waterproofing 41
        III.2.2 Expansion joints 42
                Large joints 43
                Small joints 43
        III.2.3 Bearings 46
        III.2.4 Surface and subsurface water drainage 47
        III.2.5 Repair of concrete 47
                Surface repairs 47
                Internal repairs 51
        III.2.6 Maintenance of metal structures 53
                Fatigue crack repair 53
                Painting of metal components 53
        III.2.7 Foundations 55
                Review of erosion problems 55
                Control of erosion effects and protective measures 56
                Maintenance and repair of foundations 57
                Problems caused by sediments and debris 59
                Problems caused by ice 60
    III.3 Special Techniques for Maintenance of Masonry Bridges and Timber
          Bridges 60
        III.3.1 Masonry bridges 61
                Typical defects of masonry bridges 61
                Maintenance operations 62
        III.3.2 Timber bridges 66
                Agents affecting durability 66
                Weathering 68
                Maintenance Operations: 69
                A. Maintenance measures for deck slabs and surface layers 69
                B. Maintenance measures for truss structures 72
                C. Maintenance measures on substructures 72
    III.4 Auxiliary Equipment and Techniques 72
    III.5 Conclusions 74
    Bibliography 75

Chapter IV

    MAINTENANCE POLICY - PROPOSALS 79
    IV.1 Framework of Management 79

IV.1.1	General conception of maintenance	79
IV.1.2	Management and principles of organisation	79
	Assignment of responsibility between different authorities	79
	Finance of maintenance	80
	Organisation of maintenance	80

IV.2 Maintenance Policy 80
    IV.2.1 Elaboration of a maintenance policy 81
    IV.2.2 Safety aspects 82
    IV.2.3 Economic aspects 84
           Evaluation of construction and maintenance policies 84
           Discount rates 86
           Use of discounting procedures 87
    IV.2.4 Traffic aspects 87
    IV.2.5 Technical aspects 88

IV.3 Maintenance Engineering 89
    IV.3.1 The significance of bridge design 89
           Accessibility 89
           Maintenance considerations in bridge design 89
           Replacement of deteriorated elements 90
    IV.3.2 Technical assessment for maintenance 91
    IV.3.3 Maintenance information systems 91
    IV.3.4 Training and Research 92

Reference 93

Chapter V

MAINTENANCE RESEARCH 94

V.1 Research Considerations 94
V.2 Current Research 94
V.3 Planned Research 95
V.4 Recommendations for Future Research 95
V.5 List of Some Ongoing Research 97
V.6 Bibliography 98

Chapter VI

CONCLUSIONS AND RECOMMENDATIONS 102

VI.1 Introduction 102
VI.2 Maintenance Policy - Present Situation 102
VI.3 Elaboration of a Maintenance Policy - General Principles 103
VI.4 Bridges and their Evolution in Time 104
    VI.4.1 Rates of bridge replacement 104
    VI.4.2 Design of bridges with a view to maintenance 104
    VI.4.3 Arrangements to be made when putting into service 105
    VI.4.4 Arrangements to be made during the life of the structure 105
VI.5 Organisation 105
    VI.5.1 Responsibility for maintenance 105
    VI.5.2 Planning of maintenance 106
    VI.5.3 Conduct of maintenance operations 107

VI.6 Means of Action 107
    VI.6.1 Personnel 107
    VI.6.2 Level of expenditure 107
    VI.6.3 Documentation - data bank 108
    VI.6.4 Maintenance equipment and techniques 108
VI.7 General Conclusion 109

Annex A
    EXTRACTS FROM: "DEFAUTS APPARENTS DES OUVRAGES D'ART EN BETON" 110
Annex B
    DATA ON THE REPLACEMENT OF BRIDGES IN FRANCE DURING 1978 116
Annex C
    EXAMPLE OF A BRIDGE DATA BANK SYSTEM 116
Annex D
    EXAMPLE ILLUSTRATING CONSEQUENCES OF VARYING DISCOUNT RATES 124
List of Members of the Group 127
List of Road Research Publications 130

Chapter I

INTRODUCTION

I.1 PLACE OF THE STUDY IN THE FRAMEWORK OF BRIDGE MANAGEMENT

Bridges are key elements of the road network designed to ensure efficient movement of people and goods. Their maintenance constitutes only one of the aspects of the general policy of the management of these structures from their entry into service until their reconstruction. Bridge management comprises:

1. Inspection and documentation,
2. Maintenance,
3. Evaluation of load carrying capacity,
4. Repair and strengthening,
5. Replacement and reconstruction.

Two OECD Road Research Groups have already looked into the first and third of these areas. In studying the second topic, which falls quite naturally within the general research effort on existing road bridges, it is thus appropriate to consider the various problems relating to bridge maintenance by bearing in mind the evident links with the other fields of interest. Indeed, as pointed out in this report, it is difficult to separate out the maintenance component in the overall context of management, not only for economic reasons but also in determining an overall maintenance policy or assessing the type of maintenance operation needed for a given bridge.

Lastly, it should be borne in mind that the design of bridge structures may to a large extent affect their management and, conversely, experience in bridge management should serve to refine design methods.

I.2 TERMS OF REFERENCE

The Group's terms of reference which are in line with earlier studies on inspection and evaluation of load-carrying capacity are set out below:

Background

Road bridges are designed to provide services over a long period, at least a hundred years or so. However, the deterioration of road bridges due to ageing and increasingly heavy traffic loads (and in some countries climatic conditions) is a common feature in many countries and leads to serious problems such as:

- loss of user comfort and, hence, of traffic safety;
- in certain cases, reduction of structural safety and, in some exceptional cases, risk of collapse and necessity for bridge closure;
- expenditure of large sums for repairs which could be avoided if reasonable funds were devoted to maintenance.

Bridge inspection, increasingly carried out in recent years in several countries, has revealed a large number of defects and has highlighted the need for establishing a more systematic bridge maintenance policy.

The purpose of maintenance is to keep bridges in serviceable condition in the interest of both public safety and economy. It should cover the load-bearing components of the structure but also the non-load-bearing elements sometimes regarded as secondary: pavement joints, crash barriers, etc.

<u>Tasks of the Group</u>

Within the above-mentioned context, the Group should prepare a report on the following topics:

1. Methodology for the utilisation of data provided by bridge inspection. Inventory of different forms of degradation.
2. Comparison of maintenance operations carried out at the present time in different countries and assessment of the present situation as regards bridge maintenance. Indication of the possible consequences of a lack of maintenance as regards safety and financial implication.
3. Inventory of different methods and techniques of bridge maintenance.
4. Definition of the goals of general bridge maintenance policy (goals to be reached). Indication of the means to be employed.
5. Definition of research to be carried out in order to improve road bridge maintenance techniques.

The Group's study will lay emphasis on tasks 1 to 3 and will take into account earlier OECD work on "Bridge Inspection" and "Evaluation of Load Carrying Capacity of Existing Bridges". The study will deal with all bridge types; the special case of the maintenance of large bridges will be excluded.

I.3 STATEMENT OF THE PROBLEM

The road network which is essential to the economic life of a country has been constructed to ensure continuous use by the different categories of road users in reasonable conditions of safety. This in itself justifies the need to manage, operate and hence maintain the road network in order to keep up a standard of service appropriate to traffic requirements. The maintenance of bridges, which are specific elements of the network, though only one of the aspects of bridge management policy (cf. I.1 above), should, therefore, be viewed in the context of the maintenance policy of the total network, bearing in mind the specific characteristics of bridges.

The main aspects to be taken into account in determining a maintenance policy are the following:

- bridges are sensitive points in the road network because of their strategic location and the serious consequences entailed by their failure or closure;
- in many cases bridge maintenance (and more generally bridge management) is a difficult art requiring good engineering knowledge and involving a wide range of techniques including advanced technologies;
- the diversity of structures is very important because of the many types of bridges in existence, their age, state of repair, changing regulations and construction techniques;

- the older bridges are generally not suited to present-day traffic which is characterised not only by increasing loads (heavy vehicles, exceptional convoys, etc.), but also traffic operation techniques that are potentially damaging (such as the use of de-icing salts to ensure free traffic in winter time);
- lastly, improved bridge design methods, greater knowledge about materials' behaviour and improved construction techniques have led to the acceptance of a general reduction in dimensions of modern bridge structures.

I.4 GENERAL MAINTENANCE OBJECTIVES

The general maintenance objectives have been referred to in the earlier OECD studies on existing bridges, in particular in the report on bridge inspection, an activity which must clearly precede any decision as regards maintenance. These general objectives are recounted briefly as follows (for more detail see also II.1 and IV.2):

I.4.1 Avoiding damage or injury to third parties for which the bridge authority may be liable.

Users of the bridge, whether in a direct or indirect capacity (the latter case includes, for example, those making use of a public utilities conduit placed in the structure), are not willing to accept any risk of bridge failure nor its consequences, even though it may in practice be technically or economically impossible to fully satisfy this demand. It is thus necessary to find an acceptable compromise at the policy level between this requirement and the means available, in order to ensure an appropriate level of safety.

I.4.2 Ensuring the best possible conditions for traffic

This is an essential objective since it is the very purpose for which a road network is built. Indeed, it is important to avoid any conditions leading to traffic limitations or the closure of the bridge, since the social cost is generally very high. It is also vital to avoid unscheduled interventions, which may be caused by lack of maintenance and which are contrary to the principle of good management.

I.4.3 Preserving the national bridge stock as effectively as possible

Bridges represent a considerable capital asset not only because of the heavy investment required in replacing them before due time, but also because some of them form part of the historic and cultural heritage of a country.

None of them are endowed with an eternal lifespan, much as one might wish this to be the case, and it is vital to be vigilant in safeguarding this heritage, while at the same time seeking to keep down costs especially in view of the prevailing economic difficulties. It should be emphasised that lack of maintenance generally results in reduced life span and deterioration in the bridge structure which often entails considerably increased maintenance or repair costs.

I.5 SOME ASPECTS OF BRIDGE MAINTENANCE

I.5.1 Overview of the problems

It was mentioned in I.1 that maintenance is only one aspect of bridge management. Indeed, in practice it is neither possible nor desirable to isolate this activity from others (inspection, repair, etc.) and often it is difficult to choose from among the various options. For example, assuming that user safety is ensured, is it more opportune to inspect and plan for early reconstruction or to carry out regular maintenance and repairs, bearing in mind the temporary inconvenience to traffic, the rise in costs consequent upon a deterioration in the state of the bridge, the amount of available funds, etc.?

Bridge design itself may have a marked impact on maintenance problems in regard to such factors as the choice of materials, the design of certain components and appropriate construction techniques. This idea is discussed in more detail in Section IV.3 of the report.

I.5.2 Durability

It is generally accepted that bridge life should be long, somewhere in the region of a hundred years. An examination of the age of existing bridges would appear to confirm this point of view despite considerable changes in traffic characteristics and in the technological, economic and political outlook. One would, of course, be hard put to assert that such a life span is an optimum, particularly in view of the present economic situation, but various considerations militate in favour of bridges designed for a long life span:

- some measure of overdesign (and hence slightly higher construction costs) can improve the robustness of a bridge and hence its life span;
- replacing a bridge causes great inconvenience to users which is intensified in the case of heavily trafficked bridges or those in urban areas. Such inconvenience would be intolerable were life spans to be short;
- the cost of reconstructing bridges in service is considerably higher than that of initial construction (1.5 to 8 times as much when the costs of temporary structures, the consequences for users, etc. are included).

In conclusion, the optimum life span of bridges should generally be long and depends on such factors as: ease of strengthening or replacement, the intensity of the traffic likely to be disrupted and the changes in traffic that the bridge is expected to cater for - e.g. in the light of the development of the locality.

I.5.3 Preventive maintenance

The adage "prevention is better than cure" is eminently true for bridges where defects can rapidly have serious consequences if action is not taken. As a general rule, regular planned maintenance leads to an optimisation of costs, though this rule may have to be modified in the case of certain structures where interventions may result in serious disruption of traffic. It may then be more economical to reduce the number of interventions and postpone certain maintenance operations.

I.5.4 Problems posed by the economic approach

Clearly any policy adopted should ensure traffic safety while at the same time aiming to optimise bridge management and operational costs.

This optimisation is difficult, even when deciding on the funds for immediate or short-term work, because one cannot always assess correctly the effects of the interactions between the various management activities, each of which implies different options. The overall economic situation also imposes various constraints. When it comes to optimising costs over the whole life span of the bridge, the task becomes a near impossibility as our knowledge of long-term needs and possibilities is practically zero.

To summarise, the economic approach must be considered as an aid to decision-making and not as an end in itself. Properly used, it helps in making a proper choice between alternatives. For instance, the choice between preventive maintenance and delayed maintenance could be justified on an economic basis (see IV.2.3).

For instance, certain hypothetical policy options, while economically valid from a theoretical point of view, may prove impossible in practice. Thus a deliberate policy of zero maintenance in cases where it is deemed more economic to replace a bridge when it has reached the end of its life span instead of continuing systematic maintenance must, for reasons of public safety, be accompanied by a massive stepping up in inspection and by a carefully planned programme for renewal. It could well be that the means required for such a policy are impossible to mobilise.

I.5.5 Technical difficulties and uncertainties

Generally speaking, bridge engineering problems often call for specialist knowledge, e.g. when using advanced technologies such as radiography. Furthermore, while the rapid advances in technology (high-strength steels, special concretes, cantilevered or outward thrust construction) have led to greater productivity, quite commonly a number of "teething problems" and imperfections emerge whose long-term impact may be difficult to assess.

Undeniably there has been much progress in achieving an understanding of the problems raised and a rapid development in the techniques made available for bridge inspection, maintenance and repair. It is also true that, on the basis of existing observations, further considerable improvements are possible. However, in view of the diversity of the field concerned many uncertainties remain: is the condition of the structure properly known? Have certain disquieting signs escaped us through ignorance of the phenomenon in question or the impossibility of measuring it? What is the birdge's true load-carrying capacity? How will it react over time to the wear and tear of traffic, climatic conditions, operating conditions? What is the long-term impact of such phenomena on the bridge?

Much remains to be done in regard to the implementation of a coherent policy, in view of the importance of what is at stake and the complexity of the problems involved, and to the development of our knowledge.

I.6 SCOPE OF THE STUDY

There are two distinct aspects to be considered: the definition of the field of maintenance and the nature and characteristics of the bridges concerned.

I.6.1 Definition of the field of bridge maintenance

Maintenance includes all operations designed to maintain a bridge in a serviceable condition. Serviceability is defined both in regard to the safety and comfort of users and to the life of the bridge in the conditions of use for which it was designed.

When attempting to characterise maintenance work the following typical features are noteworthy:

- firstly, the purpose of the operations to be carried out which is to safeguard the integrity of the structure and preserve it against deterioration;
- secondly, the recurrent nature of the operations, i.e., periodic work which is often of some importance (e.g., painting, replacement of joints and bearings and, more generally, the replacement of all parts having a limited life);
- lastly, the small size of the task or of the resources needed to remedy defects of accidental or chance origin.

Such is the definition of maintenance in the strict sense of the term.

In part, this report has had to take into account a wider field covering all or some of the activities related to the management of a country's stock of bridges. This is for two main reasons:

- in regard to maintenance policy, especially in its economic aspects, the various activities constituting bridge management overlap and it is impossible to isolate the maintenance activity as such. Maintenance policy should thus be integrated into a more general policy for the management of bridges and, in some cases, of the road network (cf. I.1 and I.3);
- in regard to techniques, it should be noted that some of them may be used for both maintenance and strengthening and thus cannot be excluded. Similarly, in the field of research, it is not possible to draw a clear line between maintenance and repair.

I.6.2 Nature and characteristics of the bridges concerned

The terms of reference excluded the specific case of the maintenance of large bridges. Accordingly, the study has been limited as follows:

Constituent materials

Only the most commonly used materials are considered, namely:

- reinforced concrete,
- pre-stressed concrete,
- iron and steel,
- masonry and brickwork,
- timber

Size

There is no commonly accepted lower limit though some countries have recognised the need to set one in order to draw a clear distinction between bridges and drainage structures.

No upper limit has been set but the specific problems relating to the size of a structure or parts of a structure (e.g., the technical problems posed by the jacking of large bridge decks) have not been considered. In principle, the problems raised by the maintenance of large bridges and the relevant techniques are not, with the above reservation, fundamentally different from those of small and medium-size structures.

Types of bridges

No types of structure were excluded for any a priori reason and the only bridges not included in the study were exceptional structures such as suspension and cable-stayed bridges (where the maintenance of cables is highly specialised) and structures or components whose maintenance requires techniques other than those used in civil engineering (e.g. the equipment of movable bridges).

To sum up, the study covers the large majority of bridges and the only elements excluded are those demanding highly specialised techniques.

I.7 STRUCTURE OF THE REPORT

The different chapters of the report are outlined below:

Chapter II: "Maintenance policy - Present Situation"

This chapter has been drawn up on the basis of information gathered from participating countries. It recalls the main objectives and aspects of maintenance and goes on to identify the various categories of maintenance, the criteria used to warrant their inclusion and the organisational arrangements enabling a policy to be implemented in this area. Lastly it provides some interesting information on the present rate of bridge renewal and the share of resources allotted to maintenance.

Chapter III: "Maintenance Methods and Techniques - Current Situation"

The first part of this chapter consists of an inventory of maintenance operations and a summary table on the interventions required on various elements of the structures. The second part reviews the current situation as regards certain specialised techniques (repairing concrete bridges for example) and deals with the problems and techniques involved in maintaining major bridge components (e.g. foundations) or special structures (timber bridges, masonry or brick works).

Chapter IV: "Maintenance Policy - Proposals"

This important chapter starts out by placing bridge maintenance in an overall context and goes on to deal with the problems and aspects to be taken into account in planning and implementing a maintenance policy:

- organisational and management problems;
- the main aspects to be taken into account: safety, economic, traffic and technical issues.

The last part of the chapter deals in some detail with several major policy considerations: the influence of design, the use of data banks, training and research.

Chapter V: "Maintenance Research"

Apart from identifying the main areas for research, the chapter also sets out the state-of-the-art on bridge maintenance (research published or planned).

Chapter VI: "Conclusions and Recommendations"

This chapter sums up the report's main findings: it is intended to assist decision-makers involved either at the technical or policy level in bridge management.

Chapter II

MAINTENANCE POLICY – PRESENT SITUATION

II.1 GENERAL CONCEPT OF MAINTENANCE POLICY

The work of maintaining highway bridges is usually the responsibility of local highway authorities. Even for bridges on state road and motorway networks, maintenance is often devolved to these authorities. Within a country there may be considerable variations in the organisation, practice and cost of maintenance between these authorities but centrally collected data on the work is often lacking. Information on costs is particularly difficult to obtain since bridge maintenance is usually part of the general highway maintenance budget and is often not accounted for separately. While members of the Research Group have endeavoured to provide a realistic account of bridge maintenance in their respective countries, the review of present maintenance practice in this chapter may not be complete in all the local variations.

II.1.1 Aims of bridge maintenance

The aims of bridge maintenance in various countries are summarised in Table II.1. The various aspects of maintenance policy are determined by such objectives and constraints as:

- preserving serviceability and load-carrying capacity for as long as possible;
- achieving economy as regards present and future costs;
- ensuring continued serviceability within the limits of available funds;
- assuring the safety of road users;
- minimising interference with traffic;
- providing adequate rideability and travel comfort.

The overall safety of road users would probably be served better by expenditure on the functional aspects of road safety rather than on the structural safety of bridges (i.e. avoidance of the limit state of collapse). However the public expects that bridges and viaducts will be maintained in a safe structural condition and the engineer has usually a legal obligation and always a moral obligation to ensure that this is done. The preservation of the bridge against risk of collapse is therefore one of the main aims of maintenance.

Economy in maintenance is of major importance. The objective is the long-term maintenance of the load-carrying capacity of the structure but it should be noted that, particularly in North America, bridges tend to become functionally obsolete before they are structurally obsolete. The very low rate of replacement of bridges in most countries (see II.5) shows that long-term maintenance must generally be the rule but, in some countries, the expenditure on maintenance is considered to be insufficient to prevent an excessive rate of deterioration (see II.6).

Table II.1

SUMMARY OF THE AIMS OF BRIDGE MAINTENANCE

<u>Belgium</u>

The principal objective of maintenance is to preserve, for as long as possible or desirable, the serviceability of the bridges for the loadings foreseen.

<u>Denmark</u>

To maintain to a standard commensurate with sound economics because lack of maintenance will lead to expensive repairs and replacements. Efforts are made to minimise interference with traffic since even minor restrictions cause complaints and have a psychological effect on road users.

<u>Finland</u>

To assure the safety of traffic and the serviceability of the road, to ensure as long a life span as possible with reasonable costs and to preserve the aesthetic aspects of the bridge.

<u>France</u>

To ensure that bridges will provide, permanently and with full security, the services for which they were constructed. It is therefore necessary to assure public safety by carrying out, in good time and in an economic manner, the work of safeguarding, maintaining or repairing which will allow the structure to be preserved in a serviceable condition.

<u>Germany</u>

Bridge maintenance should be considered as integral part of road transportation systems maintenance. The main aims are to ensure continued serviceability, reliability and safety, and to preserve the structures to ensure the most economical use of road transportations investments.

<u>Italy</u>

For new structures the aim is to preserve them in their original condition; for older structures the aim is to preserve them by modern techniques and materials and by the installation of new bearings, waterproofing systems, etc.

<u>Japan</u>

To assess the safety of older bridges, to repair earthquake damage and to consider environmental problems, especially traffic noise.

<u>Netherlands</u>

To prevent, to ascertain and to repair all damage and defects. This should ensure the reliability of all parts of the structure, traffic safety and a satisfactory life span.

<u>Norway</u>

Bridge maintenance is an important and often pivotal element in an integrated administrative and technical system whose overall aim is to optimise road transport safety and economy. The main aims are to protect the public from any disastrous collapse, to provide the public with safe and comfortable travel and to provide the transport industry with bridges of full design strength.

<u>Spain</u>

To maintain the bridge in good condition, equal or similar at least to the construction-design condition, with special regard to the safety of users and a long life span.

<u>Sweden</u>

To maintain carefully the comparatively large bridges needed in Sweden to ensure satisfactory operational safety and an economic service life.

Switzerland

Generally to maintain the bridges in a serviceable condition. Particular tasks requiring attention are described.

United Kingdom

To maintain highway structures in a serviceable condition with particular regard to the safety of users, the preservation and integrity of the structure and the achievement of economy in the long term.

United States

To ensure continued serviceability, adequate reliability, safety features and maintenance of design load capacity within the limits of available maintenance funds.

II.1.2 Definitions of maintenance

In this chapter maintenance is defined as the work needed to preserve the intended load-carrying capacity of the bridge and to ensure the continued safety of road users. It excludes any work leading to betterment of the structure, whether by strengthening to carry heavier loads, by widening or by vertical realignment of the road surface. It also excludes any corrective measures which are the responsibility of the contractor building the bridge, any maintenance or repair of damage due to utility services (gas, electricity, water, etc.), any repairs for which the cost can be recovered by claims on insurance and repair of any damage caused by landslide, earthquake, typhoon, fire and other exceptional causes.

II.1.3 Preventive maintenance

Where the need is foreseen for remedial work to prevent deterioration or the development of defects, the work is usually described as preventive maintenance. Whenever possible the work is done promptly as soon as any incipient defects or conditions which may lead to defects are detected. By this means more extensive repairs or replacements can often be avoided or delayed, thus reducing the cost of maintaining the structure. The type of work needed will often be influenced by the climate; examples of preventive maintenance are the impregnation or sealing of concrete to reduce the risk of frost damage or corrosion of the reinforcement, the corrosion protection of steel bridges and the prompt installation of remedial waterproofing or drainage when inspection has detected water leakage through the structure.

Regular and effective preventive maintenance is the aim of good management since overall expenditure will usually be minimised but funds are often insufficient to allow this aim to be achieved. If, through lack of preventive maintenance, major repairs or rehabilitation of a bridge become a necessity, the cost is likely to be high. In Italy, it is found that the cost of rehabilitating one bridge having deep-seated deterioration approaches the cost of reconstruction and is the same as that for the normal maintenance of ten bridges. The work of maintaining a bridge will often interfere with the flow of traffic and create a temporary hazard for road users. In some circumstances this is an important factor in determining maintenance policy.

Some doubt was expressed by France as to whether preventive maintenance of short- and medium-span bridges is achieving its objective. The preventive maintenance work done is in some cases insufficient to avoid the risk of serious deterioration and this will eventually have to be repaired at great cost.

Table II.2

AUTHORITIES RESPONSIBLE FOR THE MAINTENANCE OF BRIDGES

BELGIUM	DENMARK	FINLAND	FRANCE	GERMANY
Ministère des Travaux Publics - Bureau des Ponts - Centre de Gestion (management of maintenance) Local maintenance service	Danish Ministry of Transport - Road Directorate County Road Authorities Local Road Authorities	Roads and Waterways Administration 13 Road Districts sub-divided into 174 Road Master Districts Greater Municipalities	Ministère des Transports - Direction Départementale de l'Equipement (Routes Nationales and Motorways) D.D.E. and Departments (Departments Roads) D.D.E. and Municipalities (Local roads in large towns)	Country (Länder) Highway Departments Regional Bridge Department District Bridge Department County or City Bridge or Engineering Departments

ITALY	JAPAN	NETHERLANDS	NORWAY	SWITZERLAND
Azienda Nazionale Autonoma Strade (Ministry of Public Works) (State Road Network) - Concessionaires of motorways Provinces (Provincial Roads) Municipalities (Urban Roads)	Ministry of Construction - Japan Public Highways Corporation (National roads) Prefectures - Road Section, Dept. of Civil Engineering Districts - Tokyo Metropolitan Expressway Corp. - Hanshin Expressway Corp. etc.	Rijkswaterstat (Ministry of Transport) Provinces Towns	Statens Vegvesen Vegdirektoratet Central, Bridge Division (National Roads) County Roads Departments	Department Federal de l'Intérieur Service Cantonal des Routes - Direction générale des Travaux - Service des Routes

SPAIN	SWEDEN	UNITED KINGDOM	UNITED STATES
Ministry of Public Works Province Offices (80,000 km) Concessionaires of Toll Highways (1,200 km)	Statens Vägverk Bridge Section County Road Departments Municipalities	Transport Depts. in England, Scotland, Wales and Northern Ireland - Regional Controllers (England) (Trunk Roads & Motorways) County or Regional Councils Large Urban Authorities (Other Roads)	State Highway Department - Resident Engineer (routine maintenance in 3-5 counties) - District Engineer (non-routine work in 12-20 counties) County or City Department of Engineering

Structural maintenance is concerned with the appearance of the bridge and good workmanship will usually take care of this requirement. The repainting of steel bridges provides an opportunity to improve appearance at little extra cost. Maintenance for cosmetic reasons alone is usually of low priority except in situations where a bridge is subjected to a large amount of visual scrutiny or is part of an attractive rural landscape or urban scene.

II.1.4 Some special aspects of maintenance

Where a bridge is in poor condition through inadequate maintenance in the past, usually as a result of financial constraints, there may be a need for major restoration (or rehabilitation). This may include work needed to restore structural integrity and to correct major safety defects. It may also involve the replacement of deteriorated components.

When it has been decided that a bridge must be replaced, it is clearly desirable to reduce expenditure on maintenance to a minimum providing that this can be done without in any way conflicting with requirements of safety and traffic. Many factors will influence what can be done including the condition of the structure, the possibility of delay in its replacement and whether the existing bridge is to be retained for local use by light traffic. It is difficult to formulate criteria but good engineering judgement is essential and more frequent and thorough inspections are needed during the interim period. In Finland a system of intensified inspection allows weak, old bridges, due for replacement in 1-3 years, to be used without imposing weight restrictions. This is considered to be advantageous from the point of view of the national economy although the bridge is worn out more quickly. These principles are applied to concrete bridges, where there is thought to be sufficient warning of failure and to secondary members of steel bridges where the extent of failure would be limited.

The maintenance of large suspension and cable-stayed structures and other major bridges is not, in principle, different from that of other bridges but it may justify the employment of specialist teams and techniques. Problems of structural maintenance of large bridges are usually dealt with at a higher level in the organisation than for ordinary bridges. Inspection and maintenance manuals are sometimes provided.

The maintenance of bridges of cultural or historic interest is usually done by or in co-operation with the authority responsible for historic monuments. Some restrictions on maintenance procedures may be necessary to avoid detracting from the appearance of the bridge.

II.2 CATEGORIES OF MAINTENANCE

It is convenient to classify the many and varied tasks comprising bridge maintenance. Although the classifications used by different countries vary, probably being influenced by the manner in which the work is organised, a general pattern can be discerned. This classifies maintenance according to whether it is

- ordinary or specialised and
- periodical (i.e. programmable) or extraordinary (i.e. unpredictable).

Examples of ordinary maintenance are cleaning the bridge and the drainage systems, localised repair of surfacings, repair of traffic damage to parapets, etc.

Specialised maintenance (or repair) falls into two groups. Firstly the work for which there is, from experience, a high expectation that it will become necessary during the life span of some bridges. Examples are painting of steelwork, localised patching of concrete, replacement of joints and bearings, renewal of parts of the waterproofing or drainage system, etc. Secondly, work which is unpredictable, such as correction of settlement, renewal of post-tensioning tendons, major river training schemes, etc.

All categories of maintenance may include work needing special skills or equipment. It is often necessary to employ a specialist organisation or contractor for such work. Details of the classification are given in Chapter III.

II.3 ORGANISATION OF BRIDGE MAINTENANCE

II.3.1 Authorities responsible for bridge maintenance

Each level of the highway administration usually organises the inspection and maintenance of the bridges for which it is responsible. In most countries the organisation is as follows:

i) State highway authority, which is responsible for bridges on motorways and national routes and trunk roads. (Sometimes this authority has overall responsibility for bridge maintenance.)
ii) Counties, departments, prefectures and provinces, who are responsible for bridges on most rural and some urban roads in their areas.
iii) Municipal authorities (when responsible for the bridges in their areas).

The railways, river and harbour authorities, toll road authorities, motorway concessionnaires and private owners are usually responsible for the maintenance of bridges on their networks or property.

Few countries have a single organisation responsible for road structures and in most the authorities or agencies concerned are largely independent. In such cases the exchange of information on the solutions to problems and on specialised techniques depends more on the individual goodwill of the various organisations than upon any formal arrangements. However, many countries have one or more organisations at the forefront in the field, because of the extent or the importance of the network under their management, because of their scientific resources or because they have become the accepted authority on a particular subject. Recommendations and advice issued by these organisations provide guidance on maintenance problems for other authorities. These functions are institutionalised in some countries, for instance, the Laboratoire Central des Ponts et Chaussées in France and the Roads and Waterways Administration in Finland; for important maintenance projects, a supervisory role may be exercised.

The broad structure of the administrative system for different countries is shown in Table II.2. For simplicity, some of the regional differences have been omitted.

II.3.2 Personnel

Personnel employed on bridge maintenance includes professional or registered engineers, supervisors, foremen and skilled and unskilled maintenance workers. These may work either in units responsible only for bridge maintenance or in the general bridge or highway organisation. The work is usually planned and supervised by engineers although for some routine work this may be done by supervisors and foremen.

If maintenance is to be effective, it is usually necessary to investigate the behaviour of the structure and of the materials used. This may involve checking the design assumptions and calculations and determining the properties of the materials by tests. The workload for the engineering and supervisory staff will thus be greater than for normal construction work, particularly as maintenance is usually widely distributed geographically over a variety of types of bridge. Thus any relationship between the cost of construction and the number of personnel needed, worked out on the basis of the design and supervision of new works is not applicable to maintenance as it would result in quite inadequate staffing levels.

Resources in staff vary from large, well-equipped maintenance sections down to small units with meagre equipment suitable only for routine work. Maintenance work requires teams of 2-3 for the simpler routine tasks and up to 4-6 for specialised or major items of work.

II.3.3 Preparation and approval of the maintenance programme

Systematic inspection of highway structures is a widespread practice in most OECD countries and the maintenance programme is usually based upon these inspections. In "Bridge Inspection"[1] four types of inspection are recommended.

- Superficial inspection. This is carried out quickly and frequently by highway maintenance personnel and the purpose is to report fairly obvious deficiencies.
- General principal inspection. This is primarily a visual inspection made at 1-2 year intervals by a bridge inspector.
- Major principal inspection. This is a close examination of all parts of the structure made at intervals of 2-5 years (but as long as 10 years where there is little deterioration). It is made either by a bridge inspector working under supervision or by an engineer. Access facilities must be arranged where necessary.
- Special inspection. This is made when needed to assess the structure for exceptional loadings or to investigate major weaknesses or other problems. Special inspections may require testing and analysis of the structure; close supervision by the bridge engineer is usually necessary.

The need for maintenance, particularly emergency work, is sometimes reported by patrolmen or by the public.

The regular inspection of a large number of bridges of different types and age produces a vast amount of data and, in most OECD countries, computerised systems or card index systems are being used to obtain full advantage from both the inspection data and from past maintenance experience. Central bridge data banks have been established in some countries and these generally contain both technical and administrative data on the structures and the information available on costs. Some systems are designed to provide information directly related to the management of bridge maintenance. In such cases it is possible to obtain statistical information on the behaviour of a large number of structures with similar characteristics and this data can then be used as an aid in determining the preventive maintenance requirements for similar structures of more recent construction. Other systems provide for the recording of additional data on maintenance techniques used for various parts of the structure together with information on their subsequent performance. This enables a statistical evaluation to be made of the validity of the technique.

---

1) See list of references at the end of this Chapter.

In some data banks the information that can be stored is increased by sub-dividing the data into subjects. This also enables smaller computers to be used. A separate punched card is used for each subject and these are linked together by a code number for each bridge. Figure II.1 shows the main card and one of the subject cards (piers) for the system used in Belgium. Figure II.2 shows an example of the coding of maintenance operations in the State of Florida. Experience has shown that data must be carefully selected so that the system can be operated satisfactorily with the resources available.

The bridge inspector's reports will usually record the need for maintenance but his recommendations are normally checked by the supervising engineer who has to consider the technical aspects, the resources available, the interaction with other maintenance work and the general policy on maintenance. It is usual to prepare an annual maintenance programme but in France and elsewhere major maintenance work may be programmed over several years.

The programme of maintenance is normally approved by the authority responsible for the bridge (para. II.3.1) but, for bridges of particular importance or for any work which will require large expenditure or will be complex technically, it may be necessary to obtain approval from a higher level in the administration. Priorities may have to be established in relation to the funds available for the work and, where bridge maintenance is part of the general highway budget, allocations have to be made between bridges, roads and other items of highway expenditure.

There is usually considerable flexibility in the programme of maintenance and provision is made, sometimes from central funds, for any urgent major expenditure.

II.3.4 Carrying out the maintenance work

There is considerable variation in the manner in which maintenance work is carried out by different countries and sometimes by different districts in the same country. These differences arise from local variations in the organisation of the work and in the practices followed, but only the general trends are shown in the summary of maintenance practices given in Table II.3. This shows that ordinary maintenance is usually done by the highway departments but that most of the larger or specialised tasks are done by contractors.

For contract work, particularly where complex maintenance operations are involved, the general approach is as follows:

i) Data from bridge inspections (or at least from the major principal inspections) is collected centrally by the highway authority responsible for the bridges.
ii) Work is grouped into contracts of reasonable size. The criteria for grouping are either that the work required on several bridges is similar or that several bridges needing maintenance form a convenient geographical group. The postponement of some work can sometimes lead to a better grouping. For work needing specialised services grouping is particularly desirable.
iii) Contract work is apportioned for each item of work.
iv) The estimated costs include those for providing access to the parts of the structure where maintenance has to be done. This includes any fixed scaffolding or mobile platforms.

This approach enables a more accurate estimate of maintenance costs to be obtained and contributes to considerable overall savings. Careful planning of the work encourages firms to update their methods and provide staff training so that the quality of the work is improved.

Figure II.1 **CARDS FOR THE SYSTEM USED IN BELGIUM**

Main Card — Subject Card - Pier

BUREAU DES PONTS
SERVICE DE PROGRAMMATION, D'INFORMATION ET DE STATISTIQUE

ARCHIVAGE DES PONTS — CARTE 1

FICHE D'IDENTITE I[1]

NOM ET ADRESSE DU SERVICE

		CASES
A	CODE	1-3
B	FONCTION DE LA CARTE	4
C	N° D'IDENTITE DU PONT[2]	5-12
D	CHIFFRE DE CONTROLE	13-14
E	DATE D'ENVOI DE LA FICHE	15-20
F		21
G	DENOMINATION DU PONT	22-41
		42-61
		62-79
Z	RESERVE	80

1. A remplir en clair et à envoyer par le Service d'exécution au Bureau des Ponts lors de la décision de réaliser le pont.
2. Ce numéro est un numéro-code à indiquer par le Service d'exécution. Voir Note explicative pour les règles de sa composition.

BUREAU DES PONTS
SERVICE DE PROGRAMMATION, D'INFORMATION ET DE STATISTIQUE

ARCHIVAGE DES PONTS — CARTE 2 - PILE

FICHE D'ETUDE DETAILLEE D_i

		CASES
A	CODE	1-5
B	FONCTION DE LA CARTE	4
C	N° D'IDENTITE DU PONT	5-12
D	CHIFFRE DE CONTROLE	13-14
E	DATE D'ENVOI DE LA FICHE	15-20
F	NOMBRE TOTAL DE PILES	21-22
G	NOMBRE DE PILES DIFFERENTES	23-24
H	NOMBRE DE PILES IDENTIQUES A CELLE DECRITE CI-DESSOUS	25-26
I		27
FONDATIONS		
J	DIAMETRE (cm) ET NOMBRE DE PILES DECRITES ET INCLINEES	28-33
K	NOMBRE ET LONGUEUR (dm) DES ELEM. DROITS	34-38
L	NOMBRE ET INCLINAISON DES ELEMENTS INCLINES	39-42
		43-46
		47-49
SEMELLE		
M	LONGUEUR (dm)	50-52
N	LARGEUR (dm)	53-55
O	HAUTEUR MAXIMALE (cm)	56-58
VOILE OU COLONNES		
P	NOMBRE PAR PILE	59-60
Q	FORME	61-64
R	HAUTEUR (dm)	65-67
S	LONGUEUR (dm)	68-70
T	DIAMETRE (cm) OU EPAISSEUR (cm)	71-73
CHEVETRE OU DALLE CHAMPIGNON		
U	FORME	74
V	HAUTEUR (dm)	75-76
W	LARGEUR (cm)	77-79
Z	RESERVE	80

	CODIFICATION	VERIFICATION	PERFORATION	VERIFICATION
NOM				
DATE				

Figure II.2 EXAMPLE OF CODING OF MAINTENANCE OPERATIONS, USED BY THE STATE OF FLORIDA

PROCEDURES
State of Florida Department of Transportation

No.: 124-001   Page 5

Table 3   **FUNCTION NUMBERS**

CO.	SECTION	JOB	FUNDS	ACCOUNT			COST CENTER
				CONTROL	D	FUNCTION	
X X	0 0 0		X X X	4 3 7 1	X	X X X	X X X

ROUTINE MAINTENANCE - BRIDGES

Bridge Deck Maintenance and Repair

- 800 Expansion Dam Repair - I
- 801 Expansion Dam Repair - II
- 802 Expansion Joint Seal (Compression Elastomeric)
- 803 Expansion Joint Seal (Compression Elastomeric) Asphalt Overlay
- 804 Joint Sealant Repair
- 814 Other Bridge Deck Maintenance

Superstructure Maintenance and Repair

- 815 Beam Saddle
- 816 Timber Stringer Replacement
- 817 Painting Structural Steel - Inorganic Zinc
- 818 Painting Structural Steel - Oil Base Paint
- 829 Other Superstructure Maintenance

Substructure Maintenance and Repair

- 830 Concrete Cap Extension
- 831 Timber Cap Replacement
- 832 Timber Cap Scabs
- 833 Treated Timber Pile Replacement
- 834 Timber Pile Splice
- 835 Timber Pile Sway Bracing
- 836 Shimming Timber Piles
- 837 Steel H Pile Repair
- 838 Concrete Pile Jacket (Reinforced)
- 839 Integral Pile Jacket (Concrete) Type I, II, III, IV, V, VI
- 840 Integral Pile Jacket (Steel)
- 841 Timber Helper Bent
- 842 Helper Bent
- 843 Crutch Bent
- 849 Other Substructure Maintenance and Repair

Channel Maintenance

- 859 Other Channel Maintenance

Movable Bridge Maintenance and Repair

- 860 Routine Movable Bridge Mechanical and Electrical Maintenance
- 869 Other Movable Bridge Maintenance

Ferry Maintenance

- 897 Ferry Maintenance and Repair

Tunnel Maintenance

- 898 Tunnel Maintenance and Repair

Fishing Walk Maintenance

- 899 Fishing Walk Maintenance and Repair

Bridge Inspection

- 901 Field Phase
- 902 Office Phase
- 903 Underwater Field Phase
- 904 Underwater Office Phase
- 905 Overhead Sign Structure
- 919 Other Bridge Inspection

Bridge Engineering

- 921 Plans Preparation
- 922 Load Rating Determination
- 923 Data Management
- 929 Other Engineering

Operating Costs

- 931 Bridge Operation
- 932 Tunnel Operation
- 933 Ferry Operation

Special Accounts

- 934 Palm Valley Bridge
- 935 Special Bridge Account
  - Tampa Electric
  - Steinhatchee
  - Hutchinson Island

Any work activity that should be charged to a Job No. should be discussed with the Maintenance Engineer. The Maintenance Engineer may assign other Account Numbers not included in this manual.

II.4 ASSESSMENT OF THE NEED FOR MAINTENANCE

II.4.1 Basis for assessment

The criterion used by most countries is work needed to preserve the structural function of the bridge including the component parts and any features intended to ensure the safety of users. It is sometimes related to an evaluation of the load carrying capacity(2) made either by calculation or by loading tests. Denmark considers that maintenance is governed more by economic and traffic considerations than by structural function or load-carrying capacity. If the latter functions are in question then it is usually too late to maintain and replacement is the probable answer.

Table II.3

METHODS OF DOING MAINTENANCE WORK

Country	Highway Departments Organisation	Equipment and Facilities in Highway Department	Maintenance Work Done by Highway Department	Maintenance Work Done by Contractors or Specialist Organisations
Belgium	Management organisation for inspection and maintenance - part of Bureau des Ponts	Specialist services in the Bureau des Ponts	Minor works only	Works of importance done by contract
Denmark	Some counties have maintenance engineers, foremen and maintenance crews. Practice varies between counties but this is the trend	Maintenance crews may have spray-concrete equipment, etc. Computer facilities and testing apparatus are rented from institutions and private firms	About 50 per cent	About 50 per cent
Finland	National Board & Road Districts Control by Engineers in Construction and Maintenance Offices.(See also Table II.2) Road Master Districts: Work carried out under Road Master by Foremen, skilled and unskilled workers. Also bridge repair groups in Districts	Manual card index. A computerised data bank is being prepared. Simple testing equipment. Mobile bridge crane available. Technical Research Centre assist in advanced testing	All maintenance except painting	Painting of steel bridges
France	No separate organisation for maintenance	Card index records. Measurements or investigations are made by specially equipped laboratories	Routine or simple maintenance	Work needing major resources or particular specialisations is done by contract
Germany	Country (Länder) Departments - management, budgeting and major projects. Regional departments - contracting and supervision. District departments - routine maintenance work, smaller contracting and supervision. County or city departments similar to Country or Regional departments depending on size	Card or computerised index systems. Inspection crews with necessary equipment, smaller maintenance crews with equipment for routine maintenance	Usually routine maintenance only, occasionally smaller repair work	Most work, including major work, done by contract
Italy	A variety of administrative units are responsible for maintenance. The work is decentralised and under the control of engineers	Card index systems. Inspection equipment and specially equipped laboratories are available	Routine maintenance such as cleaning, replacement of parts, minor repairs etc.	Special works including those requiring scaffolding, the maintenance of joints and bearings, sealing of concrete, painting, etc.
Japan	No staff trained for maintenance	Card or computerised index systems. Some authorities have inspection cars, a few have simple equipment	Cleaning, but any other work is unusual	Most work done by contract
Netherlands	Maintenance Division of Rijkswaterstaat: Central Agency near The Hague (staff at M.Sc. level); Districts (covering whole country with staff at B.Sc. level)	Computerised storage of records. Special maintenance and testing equipment is rented	Cleaning of drains, gullies, etc. and supervision of maintenance work	All other work
Norway	Central Bridge Div. of Statens Vegvesen Vegdirektoratet. County Roads Dept. - Chief maintenance engineer, 1-2 Civil engineers, supervisors and maintenance teams	Index systems with construction drawings. Computer terminal	Routine maintenance. Non-routine work on smaller bridges or small-scale maintenance. Provides labour for specialist firms in remote areas	Large specialised tasks and work on larger bridges, underwater work and some replacement work
Spain	Maintenance Div. for Roads (no separate organisation for bridges)	Province Office: Maintenance Engineer Specialist services in Central Laboratory	Routine maintenance	Most work (other than routine) done by contract

Table II.3 (continued)

Country	Highway Departments Organisation	Equipment and Facilities in Highway Department	Maintenance Work Done by Highway Department	Maintenance Work Done by Contractors or Specialist Organisations
Sweden	Operating Division of Statens Vagverk, County Roads Dep.- Chief maintenance engineer, engineers and maintenance personnel	Card and computerised systems. Special maintenance equipment. Mobile bridge crane are available	Routine maintenance work and smaller repair work	Large specialised work and underwater work
Switzerland	Maintenance dealt with by Districts, each having a staff of 10	Depends upon centres for highway maintenance	Routine maintenance	Specialised maintenance by contract
United Kingdom	Counties, and some large urban authorities, usually have an engineer responsible for bridge maintenance. The size of the workforce varies. Motorway and trunk road bridges are maintained as directed by the Department of Transport's Regional Controllers	Varies between counties. Some are well equipped and supported by laboratory facilities. Card or computerised index systems are often used	Most routine work	Large or specialised maintenance tasks
United States			Routine maintenance. Other maintenance and repairs up to $50,000	Extensive work requiring large amount of manpower, equipment or materials

II.4.2 **Engineering criteria**

The criteria used to evaluate the need for maintenance includes the following:

i) The subjective assessment by an engineer with experience or training in maintenance works: most countries indicate that this forms at least the basis and sometimes the only method of assessment. For important or complex structures it may be supplemented by advice from specialists.

ii) An illustrated catalogue of defects in bridges along the lines of that used in France (3)(4) and described in Annex A.

iii) A sufficiency rating for the condition of the bridge or defined scales of assessment for defects: most countries use scales to define corrosion of steelwork.

iv) The application of testing procedures: although testing does not appear to play a major part in assessment it can be important in detecting some deficiencies, faults or the presence of deleterious materials. Examples are substandard protection against corrosion, chlorides in concrete, substandard concrete and deficiencies in important steel members.

v) An assessment of whether effective maintenance can be carried out: France, the Netherlands, Norway and the United Kingdom refer to the need to examine the effectiveness of maintenance repairs carried out earlier. There appears to be no systematic examination of the general effectiveness of maintenance.

vi) An assessment of the consequences of not doing maintenance work: Denmark, Italy, the Netherlands, Norway and the United Kingdom usually consider this possibility with particular regard to structural safety and it is sometimes used as a means of demonstrating the need for maintenance work. It is also applied to bridges which are shortly to be replaced.

Both the present condition of a bridge and the rate at which this condition is changing have to be considered in relation to the stage of deterioration at which maintenance becomes necessary.

In Belgium, it is considered difficult to rationalise the evaluation of the need for maintenance by the use of well-defined criteria. In Germany, it is noted that if a subjective evaluation by the bridge inspector shows that the safety of the structure appears to be endangered, a quantitative assessment is made. Italy comments that although one or more of the above criteria are used, they cannot always be strictly applied. In Japan a manual is used to aid assessment.

II.4.3 **Financial criteria**

Bridges usually have to be maintained in a serviceable condition for a long time. This is because user demands often increase over the years and resources for replacement tend to be inadequate. Maintenance policy thus aims at ensuring that sufficient work is done each year to prevent an accumulation of work which would impose a heavy burden on future highway budgets. There is insufficient data to enable financial requirements for a satisfactory standard of maintenance to be quantified but they can be expressed qualitatively as being the allocation of sufficient funds to prevent progressive deterioration of a country's bridges by pursuing a policy of preventive maintenance and by replacing bridges which are not economic to maintain.

In Norway, the main financial criteria applied to bridge maintenance are:

i) Whether the expenditure is justified by the benefits obtained. Formal studies are not usually made (except for an extensive study of the cost effectiveness of a corrosion protection system) but data on experience are gathered in a fairly systematic manner.

ii) The level of expenditure on bridge maintenance that would trigger reconstruction or replacement of a relatively major bridge is extremely difficult to predict with any degree of validity and depends on factors such as the resources available for bridge works. However, a discounted cost of 50 per cent of the replacement cost would probably tip the scales towards rebuilding. The same would then go for a combination of i) the present-day cost of restoration, and ii) the discounted excess maintenance costs expected over and above the national average of approximately 0.6 per cent of replacement value.

iii) Discounting methods are not used systematically in comparing the cost of maintenance with the capital costs of reconstruction or replacement.

In France, expenditure is considered to be justified if the structure is thereby assured of a long life. It is difficult to define the economic justification of maintenance as a function of age and condition alone because each case must be looked at separately; for example, maintenance may depend upon whether it is feasible to do the work given the traffic congestion which would result. In Denmark, decisions on whether to maintain, reconstruct or replace are based upon economic assessments (see Chapter IV). In Italy, maintenance is considered to be successful if not less than 10 per cent of the maintained bridges are so well preserved that they do not need reconstruction. Discounting methods are not used in Finland but replacement might be considered when maintenance costs of a bridge are equal to the average annual cost plus amortisation of the investment in reconstruction.

II.5 REPLACEMENT OF BRIDGES

The data available on the replacement of bridges is given in Table II.4 which shows the annual rate of replacement in recent years. The reasons for replacement, whether due to structural deterioration, traffic growth, etc., have not generally been recorded, but in some cases the rate is known to have been limited by the funds available. If the low rates in Table II.4 for highway systems containing a large proportion of new bridges and the high rate due to replacements necessitated by a change in vehicle regulations are omitted, then the range is between 0.2 per cent and 0.6 per cent per annum, with a higher rate of 1.6 per cent per annum in Norway. Many factors influence these rates, for instance the durability of locally available materials in the older bridge (e.g. timber or stone), different age distributions and past maintenance policies. Information on bridges replaced in France during 1978 is given in Annex B.

The problem of bridge replacement in Germany is illustrated by an example from the Rhineland-Palatinate Land. The "bridge generation cycle", which is assumed to be 60 years, is shown in the form of a spiral in Figure II.3. This shows the number of bridges in existence in 1918, the number built each year between 1918 and 1947 and the number for each year between 1947 and 1977. During the period 1977-2007 the bridge building programme will comprise:

i) the construction of over 1000 new bridges (i.e. not replacements) (see Figure II.3);

ii) the replacement of bridges built between 1918 and 1947 as they reach an age of 60 years. Between 20 and 40 bridges annually will reach this age;

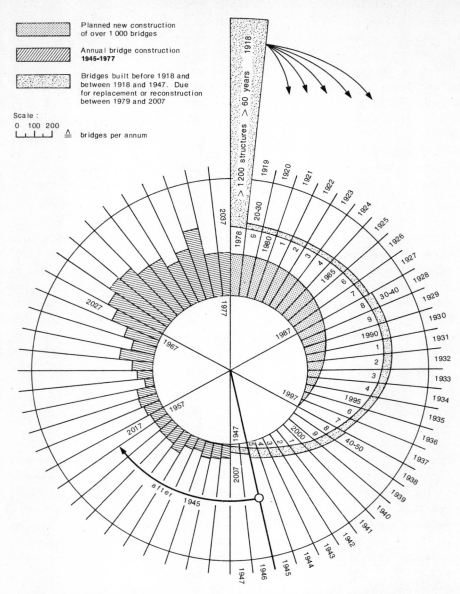

Figure II.3 BRIDGE GENERATION CYCLE = 60 YEARS

With a total future bridge stock of 8 000 structures an average bridge replacement volume is obtained of 8 000/60 that is 133 bridges p.a.

(Based on bridges in Rhineland Palatinate, Germany).

iii) the replacement or reconstruction of more than 1,200 bridges over 60 years old (built before 1918).

After 2007, the bridges built between 1947 and 1977 will progressively need replacement. Although it might be considered that a 60-year generation cycle is too short, the principle would be the same for other bridge generation cycles, even for over 100 years. The construction of bridge generation diagrams of this type would provide a useful aid to forecasting future replacements. The age distribution of bridges in Germany in another

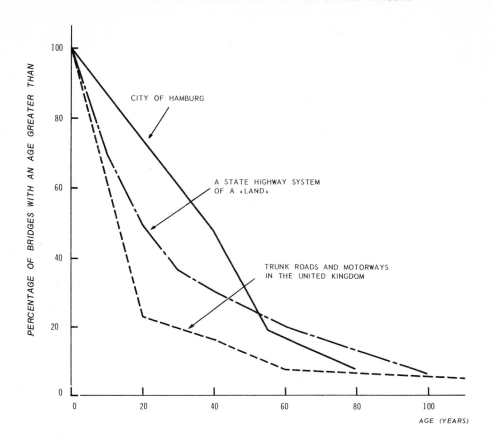

Figure II.4 **EXAMPLE OF AGE DISTRIBUTION OF BRIDGES IN TWO GERMAN «LÄNDER» AND IN THE UNITED KINGDOM**

Land and in the City of Hamburg and in the United Kingdom (trunk roads and motorways) are given in Figure II.4.

Modern conditions are likely to change the replacement rate of bridges, the main influences being:

i) The higher intensities of heavy axle loads and the widespread use of de-icing salt in colder climates (a contributory factor in corrosion and fatigue).

ii) A lessening of flood damage and erosion of foundations due to improved design and river training (thus reducing a frequent cause for the replacing of older bridges).

iii) The trend from arches and low-stressed beams to some lightweight and highly stressed modern designs may contribute to a reduction in the average life of bridges.

iv) The problems of corrosion and fatigue, particularly in the lighter structures (e.g. the life span of some post-tensioned concrete structures, may be affected by corrosion risks due to faulty grouting of the tendons).

The rates of replacement given in Table II.4 are a statement of the present position but it is doubtful whether they can give much guidance about the future. There is no

Table II.4

RATE OF BRIDGE REPLACEMENT

Country	Rate of Bridge Replacement	Rate % per annum
Belgium	No data. As first approximation assume life of 100 years	-
Denmark	2-4 bridges per 1,000 at present. Financial resources for this work are limited because they come from the general maintenance budget	0.2 to 0.4
Finland	18 bridges per 1,000 and 10 culverts per 1,000 during 1978. Part of intensive programme to replace restricted bridges following an increase in vehicle loadings in 1975	1.8
France	142 bridges per year out of approx. 50,000 (average for 1976, 77 and 78)	0.3
Germany	Overall replacement rate	0.6
Italy	5 motorway bridges out of 1,200 replaced during last 20 years	0.02*
Netherlands	1 bridge per 1,000 (due to technical obsolescence) on the State Highway system which was started in the 1930s	0.1*
Norway	16 bridges per 1,000 on National Roads (average of 1977 and 1978). 16 new and replacement bridges per 1,000 on County Roads	1.6 (National Roads)
Sweden	Estimated at 6 bridges per 1,000	0.6
United Kingdom	Estimated at 4 bridges per 1,000	0.4
United States	3,620 out of 258,000 Federal Aid bridges being replaced over a period of 7 years. Replacements limited by funds available	0.2

* These rates relate to systems containing a high proportion of new structures.

evidence which would give any assurance that rates of 0.2 to 0.4 per cent could be validly projected into the future with the implied assumptions of serviceable lives of 250 to 500 years. The problems in the United States which have reduced the serviceable life of many bridge decks to about 40 years will probably be overcome by new developments. Elsewhere, problems with some of the newer types of bridge, particularly some of the prestressed concrete bridges, are also likely to be remedied. Moreover, these types of modern structure do not require complete reconstruction because the foundations, piers and abutments can usually be retained. Thus, while some existing bridges may need replacement or reconstruction after 40-60 years, a design life of the order of 100 years is generally considered to be attainable from modern structures. Beyond 100 years, however, there must be a large element of uncertainty. An increase in the replacement rate to between 0.5 per cent and 1.0 per cent or more per annum may, therefore, have to be expected in the future.

II.6 COST OF MAINTENANCE

A summary of the information on the cost of maintenance is given in Table II.5. Where possible this is expressed in terms of the cost to the authority of replacing the bridges. It must be emphasised that the costs, which include both ordinary maintenance and the unpredictable work that is needed from time to time, are intended only to give a broad indication of the pattern of expenditure. Not too much importance should be

Table II.5

THE COST OF BRIDGE MAINTENANCE

Country	Bridges to which costs apply	Cost of maintenance	Type of maintenance	Adequacy of maintenance funds
Belgium	-	No data	Routine of preventive. Remedial	-
Denmark	1,900 bridges maintained by the Ministry of Transport Roads Directorate (excluding 200 bridges under other administrations)	0.6% per annum 1.2% per annum as percentage of replacement value	Preventive and routine (≤£5,000) incl. operating costs of a number of bascule bridges. Non-routine incl. cost of replacements	Not enough funds to prevent progressive deterioration and it is usually necessary to defer necessary work
Finland	All bridges except those in the greater municipalities. Timber bridges comprise 16% of the total	0.29% per annum (of replacement value) or $432 per bridge per annum or $2.11 per m^2 deck area per annum	Minor repairs and maintenance	Funds allocated from the general state budget to Road Districts for maintenance and construction of roads. Generally not sufficient money available to prevent progressive deterioration over the years
France	All highway bridges	0.3% per annum (of the as-new value) $10 per m^2 for suspension bridges - 1979 $5 per m^2 for metal-framed bridges - 1979	Current (routine) maintenance	After provision for indispensable highway work, the funds left for bridges are very insufficient. Priorities have to be established.
Germany	One Länder and one City (Hamburg). Average age approx. 30 years	1.0 to 1.5% 0.5 to 1.1%	All maintenance. Maintenance excluding safety measures	
Italy	1,200 bridges of the motorway concessionary company built in the last 20 years	1.5% per annum (as percentage of replacement value)	Routine and preventive maintenance	Generally satisfactory on state network. If maintenance is planned at right time, postponement is usually possible if funds are not available. On motorways work has often to be postponed because of traffic
Japan	All highway bridges	2.5% per annum	-	-
Netherlands	Bridges on State Highway System	No data	-	Until now sufficient funds have been available for maintenance but this is now beginning to change
Norway	National and County roads	0.6% per annum (of replacement value) Annual cost per bridge: National roads $1100 County roads $ 350 Average $ 700 (Bridges account for 4% of total maintenance budget)	All categories of maintenance	Sufficient money has generally been available to enable maintenance to be done at the most economic time, thus preventing progressive deterioration. It is not usually necessary to defer work for budgetary reasons. This fortunate situation is attributed to genuine concern about structural safety and the small proportion of the budget involved
Sweden	Bridges on National and County roads	Annual cost per bridge approx.$1100	All maintenance but not replacement	Sufficient funds are available for maintenance
Switzerland	-	Estimated as 0.4 to 0.8% per annum according to the quality of construction	-	Not known how effective. Cost of bridge maintenance is not accounted for separately
United Kingdom	All highway bridges	Estimated as 0.5% per annum (of replacement value). Estimated average cost per bridge $700 per annum	All maintenance but not replacement	Sufficient to cope with essential maintenance but not adequate to prevent some long-term deterioration
United States	258,000 Federal Aid bridges	$285 per bridge per annum. The proportion of the highway maintenance budget accounted for by bridges is: 6.8% on Interstate routes 4.4% on Primary routes	All maintenance	Generally sufficient funds are not available to prevent progressive deterioration over the years and it is often necessary to defer work because of limited resources. The situation is even worse at local level: there is sometimes no preventive maintenance programme

attached to differences between individual countries because these may derive from a number of factors which cannot be quantified, including the following:

  i) the types of structure, the materials used, the quality of construction, and the age distribution of the bridges;
  ii) the intensity of heavy traffic, the environment, the climate and the amount of road salt used in winter;
  iii) variations between authorities as to the items of work included as bridge maintenance and overhead costs;
  iv) differences in estimating capital or replacement values for the bridges;
  v) adequacy of past maintenance.

It is noted that a substantial proportion of maintenance costs is for traffic warning signs and control and for scaffolding or mobile platforms allowing access to the structure.

It is suggested that a reasonable interpretation of Table II.5 is that adequate maintenance requires an average annual expenditure of not less than one half of one per cent of the replacement cost of the bridge. Where the annual expenditure on maintenance is considerably less than 0.5 per cent per annum (or $700) the table shows that it is insufficient to prevent progressive deterioration. In the United States, where $285 per annum on the maintenance of Federal Aid bridges has been insufficient to prevent major problems, a special $4.2 billion 4-year programme (i.e. $16,000 per bridge) has become necessary to rehabilitate existing bridges. In Finland and France, where annual expenditures are 0.29 per cent (of replacement value) and 0.3 per cent (of as-new value) respectively, the amounts are considered insufficient to prevent deterioration. Denmark reports that annual expenditures of 0.6 per cent on preventive and routine maintenance, including the operating costs of a number of bascule bridges, and of 1.2 per cent on the larger non-routine maintenance tasks and replacements, are insufficient to prevent progressive deterioration. The bridges on which these costs were based include many major structures including some very large structures built in the 1930s on which a great deal is now being spent on repair. These costs are not expected to decrease when the repair work is completed because by then motorway bridges built in the 1960s are expected to require increasing maintenance.

In the United Kingdom maintenance expenditure of about 0.5 per cent per annum is sufficient to cope with essential work but may not be adequate to prevent long-term deterioration. In Norway, where the average traffic intensity is much lower, a similar expenditure has enabled maintenance to be done at the most economic moment. It is noted that the general trend in Norway is towards higher bridge maintenance costs in real terms. This is expected to continue until an average level of 1 per cent of the bridge replacement value is reached. In Germany maintenance costs are high but even so there is a trend for these costs to increase. Thus in Hamburg between 1967 and 1979, overall maintenance costs have increased by about 10 per cent. A similar trend can be found in Italy.

The only data on the cost of maintaining different types of bridges has been provided by Finland for timber, reinforced concrete and steel bridges. The costs, which are given in Table II.6, are for maintenance and minor repairs during the year 1978. The timber bridges were the cheapest to maintain but the table notes some special factors which may have contributed to the low costs. The relative maintenance costs of timber, reinforced concrete and steel are 0.57:0.74:1.00. In Finland maintenance costs are tending to increase. Similar information is not available elsewhere and doubts have been expressed as to whether such costs could be reliably stated in some countries because parameters such as those listed in i) to v) above will produce differing costs in structures of the same type and age.

Table II.6

COST OF MAINTENANCE AND MINOR REPAIRS TO BRIDGES
IN FINLAND DURING 1978

	Timber bridges (3)	Reinforced concrete bridges (4)	Steel bridges (4)	All bridges (2)
Number	1,526	5,518	885	8,840
Deck area (m^2)	109,806	1,228,096	253,332	1,722,170
Cost of maintenance and minor repairs ($)(1)	169,000	2,439,000	681,000	3,630,000
Cost per bridge ($)	111	442	770	412
Cost per square metre ($)	1.53	1.99	2.69	2.10

Notes: <u>General</u>: As regards timber bridges 178 are situated on main roads, 1,219 on secondary roads and about 15 are for pedestrians and cyclists. Types include: beams of solid logs (82%), glued-laminated construction (nowadays used exclusively)(14%), nailed beams (0.7%). In addition there are many wooden decks on steel bridges.

1) Overhead costs, estimated at about 30 per cent, are included in the above figures.
2) The replacement value of all bridges is $1,250 million.
3) Factors contributing to low cost include:
   a) 14 per cent are new glued-laminated timber bridges built during the past 10 years and needing very little maintenance.
   b) A large number of older timber bridges have not been maintained effectively as they are due for replacement.
   c) There is little wear from the minor local traffic normally carried.
4) There is no data to show the effect of different ages and of different amounts of traffic using the bridges.

It should be noted that expenditure on the maintenance of major bridges such as estuary crossings is not included in the data discussed above.

Bibliography

1) OECD, Road Research. Bridge Inspection. A report prepared by an OECD Road Research Group. Paris, OECD, 1976.

2) OECD, Road Research. Evaluation of Load Carrying Capacity of Existing Road Bridges. A report prepared by an OECD Road Research Group. Paris, OECD, 1979.

3) MINISTERE DE L'EQUIPEMENT. Défauts apparents des ouvrages d'art en béton, Service d'Etudes Techniques des Routes et Autoroutes, Laboratoire Central des Ponts et Chaussées, Paris, 1975.

4) MINISTERE DE L'EQUIPEMENT. Surveillance, entretien, réparation des ouvrages d'art, (S.E.R.O.70). Service d'Etudes Techniques des Routes et Autoroutes, Paris, 1977.

5) LEGER, PH. L'auscultation des ponts - Annales ITBTP No.327 - Paris, April 1975.

6) BOIS, C. L'action des laboratoires dans la surveillance et le suivi des ouvrages d'art. Bulletin de liaison des Laboratoires des Ponts et Chaussées - No.97, Paris, September-October 1978.

7) MINISTERE DES TRANSPORTS. DIRECTION GENERALE DES TRANSPORTS INTERIEURS. Instruction technique pour la surveillance et l'entretien d'ouvrages d'art. Document Direction des Routes et de la Circulation Routière, Paris, October 1979.

Chapter III

MAINTENANCE TECHNIQUES - CURRENT SITUATION

The present Chapter covers, very briefly, only the more important techniques used today in bridge maintenance. The criteria for selecting the operations to be described were as follows:

- their importance for the preservation of the structure;
- the frequency of their application; and
- their relative novelty in comparison to traditional techniques.

Detailed descriptions of well-known and established techniques are, therefore, not given. Anyone wishing to study specific subjects in further detail or to obtain additional information is referred to the bibliography at the end of this Chapter, which also contains brief summaries of the subjects dealt with in the specific publications.

Each country participating in the present study provided information on their maintenance system. The descriptions of techniques given below do not always identify the countries of origin, but refer in every instance to methods and materials currently accepted by two or more countries, thus ensuring validity and reliability of the information provided.

III.1 INVENTORY OF MAINTENANCE OPERATIONS

Before presenting such an inventory, the term "maintenance operation" needs to be specified more fully on the basis of the general concepts set out in Chapter II. There is complete agreement regarding the nature of ordinary maintenance operations, i.e. those which are classified according to the manner in which they are performed (operations of a repetitive type and, in general, technically rather simple). However, it is much more difficult to define specialised maintenance works whether these be "predictable" or "unpredictable"(1) (in certain countries they are defined as "extraordinary") insofar as the line separating these from strict repair operations is a very fine one. Several countries base the definition on the functional state of the structure, i.e. an operation is considered to be maintenance when it is performed on a structure still in service; in the contrary case it is considered to be repair. An example may serve to clarify this distinction: a bridge is a complex structure, composed of various parts, and may be considered as a "machine". By analogy with the most well-known machine, i.e. the automobile, it is easy to explain this distinction. Washing and lubricating a vehicle or performing cleaning operations on a bridge are clearly maintenance. Changing tyres on a car corresponds to the changing of the bearings on a bridge; this is also maintenance. Changing the drive shaft on a car or substituting a beam on a bridge are clearly repair operations.

---

1) "Ordinary" maintenance operations can also be divided into predictable or unpredictable work.

A specific case, rather common today, and which applies equally well to the automobile as to the bridge, is the occasional need to equip them with new devices which were either unnecessary or non-existent at the time of their construction; hence the need to protect the concrete at certain points, or to substitute expansion joints with more up-to-date and functional types, fall under the maintenance classification, defined here as "preventive" maintenance. Other examples could be mentioned, but the point should be clear that repair or replacement of certain elements can often be legitimately classed as maintenance.

An inventory of maintenance operations serves not just for the purpose of information but for programming the interventions, whether these be ordinary or specialised. In the case of the former, a simple programme can be established for the recurring maintenance operations, depending on the type of bridge structure, its operating conditions (traffic, climate, etc.). The specialised interventions, on the other hand, should be based on the results of periodic inspections to identify defects in the structures. It thus seems advisable (and is already the practice in France (see Annex A), and beginning in other countries) to draw up a catalogue of defects to be correlated with appropriate maintenance operations. In such catalogues the description of the defect should be sufficiently objective and clear, with indications as to its seriousness; in the case of the corresponding maintenance operations, on the other hand, a choice of alternatives should be offered, depending upon techno-economic and environmental conditions.

In the following, an attempt is made to provide a list of these maintenance operations. The list is divided into two parts (ordinary and specialised interventions) and indicates only the generic nature of the operation (further details will be provided in the subsequent section). The Summary Table III.1 included at the end of this section indicates, for each part of the structure, the types of routine maintenance that are necessary and the specialised maintenance that may be needed, with letters referring back to the list provided below.

Ordinary Maintenance Operations

A   Simple cleaning by mechanical means or by hand (of carriageways, footpaths, verges, joints, drains, gulleys, gutters, etc.); removal of foreign material such as trash or parasitic vegetation, and similar operations;

B   Substitution of deteriorated elements by removal and replacement operations (i.e. safety barriers);

C   Small restorations, repointing of masonry and brickwork, replacement of missing stones, sealing and repairs with cement or resin mortars;

D   Localised repairs to pavements and waterproofing, using bituminous materials;

E   Localised painting operations to protect against corrosion (safety barriers, inspection equipment, etc.); renewal of protective treatments on timbers;

F   Lubrication and greasing operations (bearings, machinery on movable bridges, etc.).

Specialised Maintenance Operations

G   Restoration of concrete (whether or not reinforced) structural parts, to be carried out with different techniques (simple or special cement mortars, synthetic mortars, etc.) including, if necessary, prior protection of the reinforcing bars against corrosion;

H   Restoration of brick or masonry structures;

I   Protection of concrete or masonry from degrading action by frost, salts or the atmosphere by means of painting (protective films), impregnation, etc.; disinfestation of timber structures;

J   Injection of cement grouts or thermosetting resins into cracks in brick, stone, reinforced or prestressed concrete structures;

K   Injection of cement grouts or synthetic resins (pure or with additives) into sheaths containing pre-stressing tendons;

L   Maintenance of bolts or weldings of metal structures; cleaning, greasing and substitution of wearing parts of same;

M   Anti-corrosion protection of metal structures, entailing complete stripping and repainting of part or all of the surfaces;

N   Repair or reconstruction of drainage systems (gullies, channels, collector and discharge pipes, etc.);

O   Repair or reconstruction of pavements or waterproofing of deck;

P   Repair or reconstruction (partial or total) of expansion joints, depending on their types;

Q   Maintenance of bearings by means of different operations depending on the types (repainting and graphitising, for example); setting of same, also by raising decks; substitution of devices;

R   Reclamation operations to river and sea beds to protect foundation structures (scour, flood waters, etc.);

S   Making up settlement on bridge approaches;

T   Replacement of any structural members (mainly for timber structures).

III.2 STATE-OF-THE-ART OF CERTAIN SPECIALISED TECHNIQUES

Certain maintenance operations entailing specialised techniques will now be examined in greater detail.

III.2.1 Bridge surfacing and waterproofing

The paving of the deck is governed primarily by user safety considerations; at the same time it is most important that protection of the structure by waterproofing is not neglected. Pavement quality (drainage geometry, skid resistance and surface roughness) must be carefully maintained on bridge structures, because the consequences of skidding or loss of control can be much more serious than on the open road, as regards both safety and traffic flow.

The inspection of pavements and waterproofing systems is sometimes guided by recommendations and/or directives giving a nomenclature of the defects recorded. The survey carried out by the Group showed that these types of documents exist only in a limited number of countries.

Table III.2 shows, for example, the nomenclature for defects related to bituminous bridge surfacings, recently established in Belgium on the basis of research into the mechanical behaviour of multi-layer bridge surfacings. This research showed that defects in bridge surfacings are affected by factors such as the traffic, the bond between the pavement layers themselves, the quality of laying techniques, the climatic conditions and the structure of the bridge.

Table III.1

SUMMARY TABLE OF MAINTENANCE INTERVENTIONS

Designation of Structural Part	Ordinary Maintenance	Specialised Maintenance
MAIN LOAD CARRYING MEMBERS		
1) Decks (slab and beams)		
- r.c.	A,C	G.I.J.N
- p.c.	A,C	G,I,J,K,N
- s.	A,E	L.M.N
- t.	A.B.E.	I,T
2) Arches and vaults		
- r.c.	A,C	G,J,N
- b.	A,C	H,I,J,N
- s.	A,E	L,M,N
3) Crossheads		
- r.c.	A	G.I,J
- p.c.	A	G.I.J.K
- t.	A,E	I,T
4) Piers and abutments		
- r.c.	A,C	G,I,J,N,R
- s.	A,E	L,M,N,R
- t.	A,E	I,T,R
5) Foundations		
- r.c.; b.	A	G.H.I.R
- t.	A	I,R,T
ANCILLARIES		
1) Pavements and waterproofing	D	O,S
2) Water disposal devices	A,B	N
3) Expansion joints	A,B	P
4) Sidewalks, kerbs	C	G
5) Bearings	A,F,E	Q
6) Parapets, barriers	B,C,E	B
7) Lighting systems	A,B	B

r.c. = reinforced concrete structures
p.c. = prestressed concrete structures
s. = steel structures
b. = masonry and brickwork structures
t. = timber structures

The process of inspecting bridge surfacing should include the following stages:

1. Classification of type of defect;
2. Evaluation of the risk of further deterioration;
3. Determination of probable causes;
4. Selection of suitable remedies.

In regard to the choice of remedies, it is sometimes necessary to carry out the work in two phases, applying firstly a temporary measure and then carrying out the definitive work. The choice must take into account the following factors:

1. The nature, seriousness and extent of the defect;
2. The danger which it constitutes to traffic and to the bridge surfacing itself;
3. The extent to which the repair operations will disrupt traffic and
4. The financing means available.

Table III.2

FACTORS AFFECTING THE OCCURRENCE AND
GROWTH OF DEFICIENCIES IN BRIDGE SURFACINGS (B.1)

Deficiency		Factors				
Class	Type	Traffic	Design of surfacing	Quality, manufacture, laying of materials	Climatic factors	Bridge structure
1. Blistering	Blistering	o	✶	x	✶	o
	Subsidence	x	✶	x	x	o
	Circular cracking	x	o	o	o	o
2. Slippage	Shoving					
	Subsidence					
	Cracking					
	Joint opening	o	✶	✶	✶	x
	Inadequate adhesion of surfacing					
3. Localised damage at edges of structural joints	Cracking	x	✶	o	✶	x
	Crazing					
4. Damage due to temperature effects	Transverse crack pattern	x	✶	x	✶	x
	Opening of laying joints	x	x	✶	✶	x
	Shoving	x	✶	x	✶	x
	Cracking occurring with resin-based surface treatment	o	x	✶	✶	o
5. Fatigue due to traffic	Cracking	✶	✶	✶	x	x
	Crazing	✶	✶	✶	x	x
	Longitudinal cracks in the orthotropic slab	✶	✶	✶	x	✶
6. Permanent deformation due to traffic	Rutting	✶	✶	✶	✶	o
	Depression	✶	x	✶	x	o
	Rippling	✶	✶	✶	✶	o
	Indentation	✶	x	x	✶	o
7. Bleeding	Water and laitance	x	x	o	✶	o
	Bituminous mastic	x	✶	x	✶	o
8. Loss of surfacing	Ravelling	✶	x	✶	✶	o
	Peeling - local loss of wearing course	✶	✶	✶	✶	o
	Potholes	✶	o	o	✶	o

o = no influence.
x = influencing factor.
✶ = major influencing factor.

When examining the options in greater detail, it should be borne in mind that temporary maintenance can slow down further development of the defect, but will not eliminate the cause. Such interventions are generally less costly for the moment, but require continued close surveillance and maintenance, the need for which will be identified by visual inspections.

The aim of definitive maintenance work is to correct the causes of the defect. Although more costly at the outset, the results are likely to be fully satisfactory, as long as they are based on the correct assessment of the problem. It is recommended that visual examinations be supplemented by taking samples or cores and making tests and measurements. It must also be pointed out that many operations are similar to those employed on road pavements in general (see Bib. B2), though the ultimate causes of the particular defect on a bridge may be different.

Temporary Interventions

a) Sealing

Sealing represents a temporary remedy in cases of:

1. Cracking due to slippage of surface layers;
2. Cracking due to movements of thermal origin;
3. Cracking of tension couplings of orthotropic slabs;
4. Circular cracks due to flattening of blisters;
5. Opening of laying joints due to slippage and/or thermal movements;
6. Loosening of the surfacing at concrete kerbs and joints.

Before sealing, cracks generally need to be widened and cleaned. The sealing is done by hand using a high quality product having the following characteristics: good adhesion with the bituminous material; capability of high extension without breaking; resistance to heat, once in place, and to climatic factors.

b) Planing and Scraping

This represents a temporary remedy in cases of shoving due to surface slippage or rutting. This measure can be considered definitive when rutting occurs in the thin wearing course only. In the latter case, the wearing course should be removed completely and replaced by a new layer having greater resistance to permanent deformations (Gussasphalt should be avoided except in cases where the layer immediately below is perfectly waterproof, so as to avoid the risk of blistering). These interventions are performed using a scarifier or pneumatic drill, depending upon the extent of the defect.

c) Patching

This is a temporary remedy in cases of localised deformation, crazing or depressions, and in cases of cracking at structural joints.

The preparatory operations consist in marking out a rectangular area extending at least 15 cm. in all directions beyond the damaged area, the cutting out and removal of broken or loose materials, cleaning and application of a primer or adhesive (e.g. bitumen emulsion) to the sides and bottom of the cavity.

The opening is then filled by placing the various layers having the same composition as those originally removed. Special care should be taken to ensure the compaction of the final layer, should this consist of bituminous material, so as to ensure that the riding quality is satisfactory.

d) Partial Substitution of the Wearing Course

This is a temporary measure in cases of peeling or localised loss of surface layer. The operations are the same as those described above for patching except that the wearing course must be removed in its entire thickness prior to refilling.

e) Repair of Blisters

When a blister results from a permeable layer sandwiched between two perfectly waterproof layers, the operations comprise the following:

- removal of the upper waterproof layer (e.g. Gussasphalt);
- providing appropriate drainage for the permeable layer;
- substitution of the material removed with an impermeable bituminous mix;
- sealing of all cracks in the wearing course so as to prevent infiltration of water.

When the blister develops at the level of the waterproof layer, the local repair consists in restoring the original design conditions at the interface, i.e. either adherence or independence between the waterproofing layer and the deck.

Definitive Interventions

a) Replacing the Surfacing

The purpose of this operation is to remove the causes of the deficiencies of the surfacing consisting of different layers constructed over the bridge deck. The deficiencies are assessed in relation to the initial design and on the basis of the results of visual inspection, samplings, measurements and laboratory tests. When replacing the old surfacing, it may be necessary to:

1) modify the design of the layer, e.g. removal of the cement mortar layer used to protect the waterproofing layer as this may be the cause of the deficiency;
2) use another mix design, e.g. providing a waterproofing layer of high creep resistance (see para. on waterproofing), employing a modified bituminous binder, or
3) install a drainage system within the surfacing.

It is recommended to investigate very closely the old surfacing material so as to substantiate or correct the original assessment of the defect. These observations should concentrate on the following:

1) actual crack depth;
2) age of the materials of the various layers;
3) quantity and location of infiltrations;
4) conditions at the interface between different layers;
5) adhesion of waterproofing to the deck;
6) conditions of the deck.

The new surfacing should be laid according to specifications and recommendations used for new bridges. On the basis of the results of these assessments, specifications for work contracts can be improved.

b) Restoring Skid Resistance

The purpose of this intervention is to restore the surface texture of the existing wearing course. Depending upon its nature, this can be achieved by:

1) heating the bituminous wearing course and rolling in chippings;
2) heating and cutting the Gussasphalt;
3) applying different types of surface treatments, depending on the type of support and the type of traffic on the bridge.

The surface treatment techniques comprise a wide range of possibilities, especially with respect to the type of binder used. Especially on urban bridges, where the consequences of accidents due to lack of adequate skid resistance may be particularly serious, certain countries are accustomed to use very expensive binders (generally epoxy-tars) and special aggregates such as calcined bauxite, which provide surface courses with very high friction coefficients and durability (cf. Bib. B2).

Waterproofing

Little comment is offered on this topic, which has been treated at length by a special OECD Research Group (Bib. B2) dealing with the waterproofing of concrete decks. In the context of bridge maintenance waterproofing, along with other special devices to seal the joints, is essential to ensure the durability of the structures. This is especially true for prestressed concrete structures, insofar as the presence of the waterproofing layer protects the tensioned tendons from contact with corrosive agents, especially in the case of post-tensioned cables where the sheathes may often be imperfectly filled.

Certain countries (cf. Bib. A3) also include under preventive maintenance the replacement of waterproofing layers which have lost their effectiveness, or the laying of new waterproofing layers on structures which were previously not so equipped.

In regard to the techniques for use on concrete, present trends indicate that synthetic resins are now considerably less applied because truly effective results can only be obtained by employing substantial quantities with, consequently, prohibitive costs. The most commonly applied techniques involve the use of mastic asphalts and reinforced bituminous sheeting. The former are one of the oldest types of waterproofing and they continue to be applied in traditional fashion, with slight variations (such as the use of rubber or some natural bitumens) to render the mixture more stable at high temperatures. Reinforced bituminous sheeting is a more recent development, especially the types using non-woven fabrics such as polyester as reinforcing. This type of waterproofing is especially recommended for decks where tensile stresses occur in the concrete surface (such as in continuous beams or spine beams with extended lateral overhangs) because of its resistance to cracking. These sheetings are also used frequently in localised waterproofing applications or on small bridges where other types of treatment would be less economical.

There are various application methods, depending on national practice, the weather at the time of installation and the condition of the concrete deck. The most widely used types of prefabricated sheeting are those that are flame-heated and applied over a layer of hot bitumen previously applied to the deck, which dissolves the bituminous material of the sheet. It is also possible to use an on-site construction method (Bib. B4).

According to comments from the various countries surveyed, all the different deck waterproofing systems in current use have a limited effectiveness over time. The concept of durability as applied to structures or structural components is difficult to define; waterproofing of <u>limited duration</u> is understood as having a shorter effective life than that of the structure itself, i.e. the protective layers will have to be replaced after a certain number of years, depending on the type used. Some, such as mastic asphalt on a rigid structure, may last ten or even twenty years without losing their effectiveness. It should be pointed out that in the case of prestressed concrete structures the loss of waterproofing even at small isolated points may have severe consequences. For this reason, studies have been carried out in recent years to develop more reliable systems which will have a durability similar to that of the structures themselves.

One such system is the Polymer Impregnated Concrete (P.I.C.) technique (Bib. B4). Originally developed in the United States and applied experimentally in Italy, this system consists in the impregnation of the upper part of the concrete slabs with a heat-hardening polymer to a depth of from 3 to 5 cm. This is a rather costly technique, however, insofar as the special application equipment needed is not yet available, but it could prove extremely valuable in safeguarding bridge structures.

The technique requires the complete "drying-out" of the slabs by heating to 150-180°C for a certain minimum time period; the surface is then allowed to cool down to 50-60°C and protected from rain. An overlay of heat-hardening monomer is then applied to impregnate it whereby the viscosity of the monomer is gauged to the porosity of the concrete. The new surface is re-heated after first wetting it with water to ensure that the monomer polymerises in uniform fashion (the wetting serves to distribute the heat throughout the mass of the monomer). The resulting impregnated concrete has exceptional resistance to the attacks of physical/chemical agents to which bridge slabs are exposed.

This brief description of the P.I.C. technique shows the difficulty of its application on structures under traffic. It is obviously more suitable for use during construction (especially on prefabricated components which can be treated under ideal conditions). Experiments carried out on bridges already in operation have shown the technical feasibility of this system, but further research is necessary.

III.2.2 Expansion joints

In the national replies to the Group's survey the term "expansion joint" was generally used to describe the devices used to provide trafficable zones at those locations where the bridge movements due to thermal expansion or to the traffic in general are concentrated. These devices are necessary for two main reasons:

- to permit traffic to pass over the breaks in the continuity of the structure with least disturbance;
- to protect the underlying structural parts from aggressive materials and dirt which can infiltrate at these breaks.

Of course, these devices must not in any way prevent the normal movements of the structure envisaged at these points.

From the standpoint of maintaining the bridge structure, it is important to keep in mind these two main objects and, naturally, to ensure that the device works effectively. In fact, malfunctioning or blocking of the joint may jeopardize the bridge structure itself and it is precisely because of problems of structural maintenance which have occurred in recent years that there is a growing awareness of the importance of the protective function of joints (1). Therefore, joints in modern structures generally consist of two, often separate devices, one destined to ensure the continuity of the structure (continuity joint), and the other to protect it from water and dirt infiltration (waterproof joint). The latter is often associated with drainage systems or underlying protective devices in concrete because of the difficulties often encountered in obtaining a waterproofing of sufficient long-term durability. Until recently many countries tended to overlook the importance of this function, and maintenance operations concentrated solely on the structural conditions of the joint and the elimination of possible danger or disturbance to vehicles.

It should be mentioned that from the maintenance viewpoint the operations to ensure the functioning of a continuity joint are always given urgent priority, whereas those connected with the waterproofing are likely to be postponed, insofar as the deterioration due to waterproofing defects is a rather slow process, depending on environmental conditions.

---

1) In Italy, for example, 79 per cent of special maintenance and repair operations on bridges arise due to poor sealing of the expansion joints.

In the first instance, the breakdown of a joint component, or the differences in level which develop between the pavement and the joint, are either a danger to traffic or at least will result in damage to the joint device given the continuous impacts to which it is subjected, especially under heavy traffic.

These problems are well-known in most Member countries. Several countries are now carrying out preventive maintenance by equipping non-waterproof joints with suitable protective devices or even installing completely new joints where it is not possible to provide for such protections. On the other hand, the alternative possibility, i.e. eliminating joints wherever structurally possible, should not be overlooked.

When discussing maintenance techniques in greater detail, it is necessary to distinguish between the two main types of expansion joints. First of all, there are the "large joints", i.e. where the prevailing movement is due to thermal expansion, and "small joints" present on structures where the thermal expansion does not exceed 1-2 cm, and where the more important movements are those generated by traffic.

Large Joints

These are generally of the metal type, either the more traditional comb or coverplate types, or the steel and rubber profiles inserted in varying number according to the amount of the expansion, or the vulcanised steel in neoprene type. Descriptions of these well-known types are not presented here and the reader wishing further detail is referred to the bibliography C1 and C4.

Maintenance operations for continuity joints consist of dismantling and re-assembly, cleaning and greasing, generally carried out by specialised personnel; sometimes it is also necessary to substitute certain elements such as moulded rubber elements, etc. With the more modern types these operations are facilitated by the use of modular devices which can easily be broken down into similar elements and, generally, waterproofing is ensured by strips of elastomer, fastened below the metal sections either fixed to these, or, better still, to the edges of the adjacent structures of the bridge. The maintenance to this part of the joint consists generally in cleaning operations (with jets of water sprayed from the sides or from the top of the structure), or periodic substitution of the whole rubber sheath which can easily be inserted from below (i.e. without having to dismantle the continuity joint), even in devices which were not originally designed with this in mind. It is very important that these devices should consist of materials resistant to ultraviolet decay, especially in large joints where the sunlight can more easily penetrate. Finally, it should be mentioned that such interventions also fall into the class of preventive maintenance.

Small Joints

The range of smaller joints is much more diversified; these may be similar to the large ones but constructed on a smaller scale, or, as is more frequently the case, they may simply be nosings of epoxy resin or, far less frequently, cement mortar. Finally, there are buried joints on bridges having a span not exceeding thirty metres where movements due to traffic are limited, either because of the massive nature of the structure, or the type of traffic it normally carries.

The maintenance techniques naturally vary in accordance with the type of joint. For metal types the same techniques are used as in the case of the larger ones described above. In the case of resin mortar joints, maintenance will often consist in reconstituting the nosings, or parts of them, damaged by the traffic. In fact, if the nosings are laid under unfavourable temperature or moisture conditions, they may not adhere properly to the concrete of the deck. Other difficulties can arise with the use of mortars which are sensitive to the action of ozone or ultra-violet rays (brittleness), or which have too high a thermal expansion coefficient with respect to that of the concrete.

This results in the formation of transverse cracks in the nosings and renders them easily detachable. Repairs to these are facilitated if the waterproofing is ensured by a profile of neoprene or other material bonded to the vertical edges of the concrete slab (Figure III.1).

Figure III.1 ILLUSTRATION OF THE RESIN-NOSING TYPE JOINT WITH RUBBER PROFILE TO FACILITATE MAINTENANCE

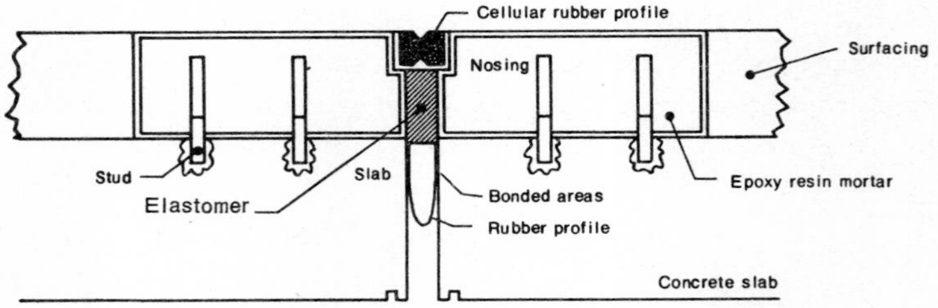

In fact, the rubber profile, besides serving to drain off the water which may infiltrate between the deck and the nosing (through cracks or limited detached areas, even if the nosings are solidly attached), permits repairs to be made to the damaged parts of the nosings without having to touch the waterproof joint. The upper rubber profile can be removed and replaced in such cases without difficulty. In replacing the nosings, or a part of them, the connection with the deck can be improved by inserting a number of steel studs into the latter (in holes made by drill or small core borers), sealing them with pure epoxy resins. There are various types of nosing joints which permit maintenance operations of this type with greater facility than is possible with modular metal joints; these consist of prefabricated nosing joints in lengths of around 1 metre. These elements are constructed in cement mortar or (rarely) in resin mortars which are simply glued to the deck. In several versions the rubber profile in contact with the traffic is also modular, which makes replacements easier in cases of accidents during winter maintenance operations (many types of joints, in fact, suffer damage when struck by snowplough blades).

Buried joints are generally the least costly to construct and, when laid on structures for which they are suited, cost less to maintain. In general they consist of waterproofing and sliding layers which waterproof the joint and spread its movements over a wider area of surfacing. Continuity of the joint is provided by the surfacing itself. In the first instance, it is preferable to pre-form the break in the pavement by means of a 3-4 cm. cut, sealed with bituminous mastic (see Figure III.2).(1)

The maintenance operations consist in periodic re-sealing covering in addition any possible cracks which might form in the winter season. The types of special surfacing most easy to maintain are those in which the paving over the joint crack consists of poured asphalt (plug-type joints - see Figure III.3 for a type used in Italy).

---

1) Special types of surfacing material of greater flexibility will reduce cracking and reduce the need for both the sliding layers and the slot.

Figure III.2 **BURIED JOINT : SEALED SLOT**

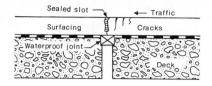

Figure III.3 **PLUG-TYPE JOINT «AUTOSTRADE»**

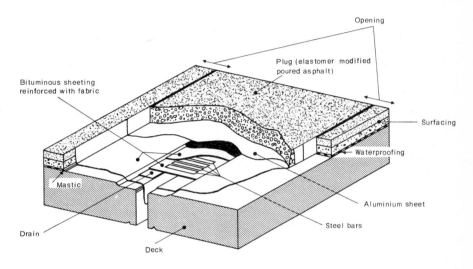

In these joints the "plug" does not follow the movements of the bridge, and the expansions (generally rather limited) occurs at the points of contact with the normal surfacing. The poured asphalt is subject only to bending at its central point due to the deformation of the deck under traffic, and holds up much better under this type of fatigue than the types of joints where the part in continuity consists of normal bituminous mixes "reinforced" with polyester webbing or fabric. The maintenance operations in the case of plug-type joints are limited to periodic resealing of the borders. Because of its low cost, the plug-type joint can also be used for preventive maintenance operations on bridges which, because of their limited span, were not originally equipped with waterproofing devices (e.g. overpasses, underpasses, etc.).

III.2.3 <u>Bearings</u>

Proper functioning of the bearings is of basic importance in the preservation of modern bridge structures. These devices, if correctly dimensioned, must ensure the distribution of loads as calculated by the designer. If maintenance of bearings and their supporting structures is neglected, the functional behaviour of the bridge may be affected and damage of variable severity may be caused including, in extreme instances, collapse of the structure. There are cases where this has occurred, even newly-built bridges have failed in this way.

In the majority of cases the location of the bearings coincides with that of the joints and, hence, the maintenance operations on the latter can also be considered as a protection of the underlying bearing equipment. For the operations to be performed directly on the bearings, it is necessary to distinguish between the different types of bearing. For traditional steel types (roller bearings, rocker bearings, etc.) the purpose is mainly to protect against corrosion and to maintain the required mobility; hence it is necessary to repaint them completely, i.e. first stripping and then repainting. After the painting, or independently of it, the greasing and graphitising operations are carried out. Very often, however, this type of bearing presents rather inconvenient maintenance problems; for example, there may be a defective alignment between the various elements (in the case of rockers) or an excessive shifting of the rollers. In these cases it is necessary, in order to restore proper functioning, to jack up the decks in order to put the elements back in place.

In the case of metal bearings, more serious difficulties often arise from an unbalanced stress distribution in the supporting concrete. To remedy these, it is possible to inject synthetic resins or cement mortars into the damaged support, to band it, or, in the most severe cases, to reconstruct it - all techniques which will be taken up in more detail in the following sections.

In recent years a major development has occurred in the use of steel or aluminium bearings with large cylindrical or spherical sliding surfaces which take advantage of the low friction properties of steel-PTFE(1) contact. These bearings, which can be built and assembled in such a way as to better ensure the mobility conditions required in the original design, also appear to be more reliable from the maintenance standpoint. This is due to both their form and the materials in contact, as well as to the protective devices with which they are generally equipped. These types of bearings have not been in use for very long periods of time, but it is expected that the maintenance operations, when and if necessary, will consist in the dismantling of the bearing and the replacement of the wearing sections, i.e. probably the PTFE layers. Of course, these operations require that the PTFE be detached from the steel backing to which it is usually bonded or mechanically fitted, while, on the other hand, rebonding, lubrication and fitting together is a job requiring clean factory conditions and new or temporary bearings will probably be used while the old ones are being repaired elsewhere.

In the case of the other types of bearings commonly used, often maintenance consists only in the substitution of the elements, e.g. fittings of neoprene or, now rarely used, lead plates. In both cases, there is complete substitution of the elements without any re-use, since the operation becomes necessary only when the constituent materials have lost their qualities of elasticity and flexibility necessary for proper functioning.

No country has set up an _a priori_ timetable for this substitution. In general, this work is carried out when inspections show that the materials have lost their functional characteristics. In the case of the old-type elastomeric bearings, i.e. consisting of simple overlays of steel and neoprene, without vulcanisation and protection at the outside borders, it has been noted that the functional life of the materials has sometimes been short. However, durability is increased with anti-ozone treatment of the rubber and by fully encasing the steel plates.

There are bearing types (pot-bearings) which combine the characteristics of those in neoprene and those in steel-teflon: the neoprene is contained in metal slide-guides and

---

1) PTFE = Polytetrafluorethylene.

functions similarly to a viscous liquid, permitting rotations of the deck. From the maintenance standpoint these types of bearings resemble those in steel-teflon with spherical or cylindrical hinges.

As can be noted, the majority of maintenance operations for bearings consist in their substitution; for this reason it is necessary to design or subsequently modify bridge structures so that the deck head-pieces are of a shape and size to facilitate their lifting for maintenance purposes.

III.2.4 <u>Surface and sub-surface water drainage</u>

There are no special techniques for these devices, insofar as their maintenance is mainly routine. It is necessary, however, to emphasize the importance of these operations. When these devices are being replaced or relaid, certain points should be kept in mind:

- it is necessary to drain not only the surface water which runs off the surfacing, but also water which infiltrates much more slowly within it; this can be achieved by paying close attention to the form of the discharge gullies, the manner in which they are laid, as well as to their number and their location;
- it is important to inspect locations where water is being drained; it quite frequently happens that the water is discharged at vulnerable points such as on the lower flange of the beams or is blown by the wind into such points.

The drainage system should be conceived in such a way as to facilitate maintenance operations; from this standpoint the standards applied in Switzerland (D1) provide a wealth of suggestions and details.

III.2.5 <u>Repair of concrete</u>

This is a common type of maintenance operation, and may be necessary at many different points on bridge structures. The difficulties and probable durability of the intervention itself can vary considerably, depending upon its size and where it is located. Until just a few years ago the only types of materials used in such works were normal cement mortars with possible addition of plasticizers or anti-shrinkage products based on metal additives. For this reason such interventions were very often carried out by ordinary construction firms, and generally with mediocre results. Today the range of possibilities is quite wide. Special cement mortars and synthetic materials offer good adherence characteristics, low shrinkage, strong mechanical resistance and low permeability, and ensure good durability. Proper use of these products, however, requires considerable experience, in order to take advantage of their high potential quality.

In the discussions hereunder, a number of alternatives are briefly described and a general assessment of their validity is given based on the experience of the countries participating in the study. A distinction is made on the basis of the type of work, i.e. depending upon whether the intervention on the concrete structure is external or internal.

<u>Surface Repairs</u>

Table III.3 presents the materials to be considered for this type of repair and indicates those not recommended. (The Table is taken from the guide published in France in 1977 - see Bibliography E2). A condition common to all these interventions is that proper preparation of the surface is essential if the intervention is to be a success. The operations are generally as follows:

Table III.3

SURFACE REPAIR IN A THICKNESS OF 2 CENTIMETRES OR MORE

		THERMO-SETTING POLYMERS	THERMO-PLASTIC POLYMERS	TRADITIONAL CEMENT BINDERS	SPECIAL CEMENT BINDERS	THERMO-SETTING POLYMERS + TRADITIONAL CEMENT BINDERS	THERMO-PLASTIC + TRADITIONAL CEMENT BINDERS
		Primer			With primer – Epoxy binder		Binding layer
Type of Usage	Binder	P	N.R.	N.R.	N.R.	N.R.	P
	Mortar	P	N.R.	P	P	P	P*
	Concrete	P	N.R.	P	P	P	P
Nature and Condition of Support	Loaded	P	N.R.	N.R.	N.R.	P	P
	Not loaded	P	N.R.	P	P	P	P
	Dry	P	N.R.	P	P	P	P
	Humid	P	N.R.	P	P	P	P
Quality of repair	Water-proofing	P	N.R.	N.R.	N.R.	P	P
	Deformation ability	Variable depending on polymers and mix design		Low	Low	Low	Better than traditional cement binders
Application at 20°C	< 12 h	P	N.R.	N.R.	P	N.R.	N.R.
	12 h to 24 h	P	N.R.	N.R.	P	N.R.	N.R.
	> 24 h	P	N.R.	P	P	P	P

P = Possible.
N.R. = Not recommended.
* If manufactured in a mixer, high voids content may occur.

- removal of the deficient concrete down to the level of the sound material;
- complete removal of all oil or grease on the concrete which could impair surface adhesion;
- removal of surface mortar and possible curing products;
- removal of rust from any exposed steel reinforcing bars, and the possible passivation of the latter (by means of anti-oxidizing products).

The deficient concrete can be removed by manual or mechanical chipping with small drills or by high-pressure blasting with water, or better, grit-blasting. This procedure also permits dressing the reinforcing steel at the same time. The prepared surfaces are given different treatments depending on the type of materials to be used. It is not possible to discuss in detail all the operations involved and only the essential points will be highlighted (i.e. the general characteristics which the products should have).

In general, it is necessary to apply a primer which serves to increase the adhesion between the support and the restoration; in the majority of cases this consists of a suitably formulated epoxy resin (or, less frequently, a polyurethane or polyester resin), consisting of a base plus a hardener. Cement mortars or even synthetic binder mortars can then be applied over the latter. These mortars should have the optimum combination of the following characteristics:

- high strength;
- good bond;
- rapid hardening;
- limited or nil shrinkage (in some circumstances a slight expanding effect is desirable);
- limited porosity.

Some of the above characteristics are necessary for the durability of the repair, whereas others are required due to the difficult conditions under which the repair is made, e.g. the presence of traffic. Indeed many countries expressed concern regarding the effects of traffic-generated vibrations on the hardening and curing processes of the repair materials. Recent studies show that vibrations have only a secondary effect on synthetic or cement mortars, but they seem to have a somewhat more unfavourable effect on sprayed concretes.

The products which appear to be in most widespread use today are mortars with epoxy binders, combined with siliceous fillers. As yet, however, there are no standards specifying their compositions and strengths for specific uses. Problems which can arise in the use of these types of mortars are linked to the extreme variability of the results which may occur if the hardener formulation is not suited to the specific purpose (use of flowing or plastifying agents which do not form a structural bond with the resin, use of products more or less compatible with moisture, etc.). In certain cases the results may be quite disappointing.

Other products in common use are "special" cement mortars which are applied by means of traditional techniques. These are generally marketed in pre-mixed form (aggregates plus binder) so as to avoid mistakes in mix designation. The ones in most common use are those containing super-plasticizers (rheo-plastic mortars) which are workable (fluid) with very low water-cement ratios.

Their strength and permeability characteristics are mainly derived from their low water content and this makes them highly sensitive to the type of curing to which they are submitted. It is necessary, therefore, to avoid the use of wooden forms (too porous) and to apply curing agents after laying. Sprayed concrete made of these products has been found to be more waterproof and durable. In many countries it is the practice to paint or protect the concrete after the above-described repair has been made. This procedure is useful in many cases if the damage occurs in concrete that is easily prone to attack by environmental agents and because it is probable that the concrete will continue to deteriorate in the unrepaired zones.

This protection may also be applied preventively to new, or so far, unrepaired structures and the techniques are generally quite similar. The treatment may vary, however, in respect to the different nature and purpose of the components to be protected, as well as their location on the bridge. It is evident, for example, that in concrete beams prestressed by post-tensioned tendons, the anchorages and the surrounding concrete require the adoption of techniques and materials which permit the best possible protection, whereas the safety margins can be reduced in the case of massive structures.

Another factor is the quality and make-up of the concretes to be protected. A porous Portland cement concrete, for example, is more prone to deterioration than a compact structure of pozzolan cement. For these reasons, then, there are generally two classes of concrete protection:

1. Maximum protection for: ends of the bridge deck and beams, end-diaphragm beams, tops of cross-heads of piers and of abutments, toppings in parapet zones (including dripstones, support zones at hinges; thin structures in general.
2. Relatively minor protection of the body of the beams and intermediate diaphragms, piers and abutments of light-weight structures.

The first category may also include overpass piers and abutments. Naturally, it can happen that under particular climatic conditions or with a greater or lesser degree of atmospheric pollution certain structures belonging to the second category will have to be given maximum protection.

Surface protection of concrete can be achieved with:(1)

- products which impregnate the concrete (whether or not reacting with the constituents more prone to attack by aggressive agents), and thus considerably reduce its surface porosity and stabilize its outer layer;
- products adhering strongly to the concrete to form a continuous highly waterproof skin, resistant to attacks by aggressive agents.

Naturally, these two types of protection can be combined. By dividing the products according to the above types of action, and also considering the main components, or at least those which tend to characterise the product itself, the following breakdown of types of protection can be proposed:

TYPES OF PROTECTION

- <u>Impregnating Products</u>
    A 1. Inorganic salts
    A 2. Dilute epoxy resins
    A 3. Other dilute resins (polyurethane, acrilic, etc.)
    A 4. Linseed oil

- <u>Products providing a protective film</u>
    B 1. Epoxy resins
    B 2. Polyurethane resins
    B 3. Chlorinated rubbers

---

1) Techniques for the surface protection of concrete are constantly developing and some of the assessments presented here may become invalid.

If the protection is to be effective, the following general requirements must be met:

- strong resistance to climatic action and to aggressive agents;
- strong resistance to ageing;
- strong adherence to support or good impregnation characteristics.

In the case of highly waterproof products, it is necessary that these do not also constitute an impenetrable barrier to evaporation, so as to avoid possible bilstering and loosening of the film, or, in the case where the product is totally impermeable to water and water vapour, that this be recognised and taken into account.

With respect to the elastic and plastic characteristics of the protection (in the case of films), it is considered that it would be very difficult to find a product of such high plasticity, in view of the thinness of the layer applied, as to be able to withstand cracking in the underlying concrete. Thus the only rheological property required is the ability to accommodate under structural expansion movements, or possibly the movements of the micro-cracks already present at the time the protective coating is applied. On the basis of the information obtained, it emerges that the best protections from the technical standpoint are as follows for the two categories:

1. For maximum protection a two-stage intervention is required; a primer coat with an impregnating product of the A2 or A3 (dilute resins) type to reduce surface porosity, fill up existing micro-cracks and prepare the surface for the second coat. Account should naturally be taken of the condition of the surfaces to be treated; these should be sound and free of impurities, small hollows, etc., so as to ensure optimum absorption of the product. A very important factor in this regard is the type of release agent used to facilitate removal of the forms in cases where newly-constructed works are to be treated. Indeed, one should avoid the use of mineral-based products which penetrate the surface to be treated and thus impede the penetration of the protective product.

   For the second stage one can use a treatment with type B2 resins. In fact, the polyurethane resins are much more resistant to ageing phenomena than the B1 and B3 products, as the former are subject to chalking and loss of their mechanical characteristics, and the latter tend to age more rapidly.

2. For this category the possible treatments vary according to the greater or lesser aggressiveness of the environment, i.e. only impregnating products with either chemical-mechanical protection (A2-A3) for greater aggressiveness or strictly chemical (A1) products for lesser aggressiveness.

Internal Repairs

With regard to the injection of cracks or cavities, Table III.4, drawn from the guide mentioned earlier, presents an inventory of repair techniques. They in no way serve to reinforce the structure, but only to prevent corrosion of the steel which could result from the free circulation of water through the cracks. Here, too, the products most commonly used in the repair of both passive (dead) cracks or active (1) ones are epoxy resins; as in all their other uses, good performance depends on the formulation of the resin and the application techniques.

---

1) However, such repairs of active cracks are seldom satisfactory and plastic sealers are more suitable.

Table III.4

INTERNAL REPAIRS - INJECTION OF CRACKS AND CAVITIES

Characteristics		THERMO-SETTING POLYMERS	THERMO-PLASTIC POLYMERS	CEMENT BINDERS Traditional	CEMENT BINDERS Special	CEMENT BINDERS Traditional with Thermo-setting polymers	CEMENT BINDERS Traditional with Thermo-plastic polymers
Aim of repairing cracks	Restore waterproofing		P	N.R.	N.R.	P	P
	Improve tensile strength	P	N.R.	P	P	N.R.	N.R.
	Improve compressive strength			P	P	P	P
Conditions of support	Dry	P		P	P	P	P
	Moist	P	P	P	P	P	P
	Under water			P with some reserve	P with some reserve		
Crack width (L)	L < 0,2 mm	P with some reserve	P	Application N.R.	Application N.R.	Application N.R.	Application N.R.
	0,2 mm ≤ L < 0,6 mm	P	P	Application N.R.	Application N.R.	Application N.R.	Application N.R.
	0,6 mm ≤ L < 3 mm	P	P	Application N.R.	Application N.R.	Application N.R.	Application N.R.
	L ≥ 3 mm	P	Application N.R.	P	P	P	P
Internal cavities		P but costly	Application N.R.	P with some reserve	P with some reserve	P	Application N.R. (decrease in strength as a function of air content)

P = Possible.   N.R. = Not recommended.

With these operations it is often impossible to provide a preventive treatment. In cracks of a certain age it is impossible to eliminate the calcite or other impurities which may form. When carrying out the injection the general practice is to insert injector tubes into the two extremities of the crack and at intermediate points (30-40 cm intervals), then to seal the crack (in cases of active cracks, the sealing is preceded by the chipping of a V-shaped channel along the length of the crack, in such a way as to provide for a greater thickness of sealing material at the most exposed point). The product injected has generally a hardener compatible with the moisture level and containing re-agent additives which form a structural bond between the resin and the concrete. The viscosity of the formulation should be compatible with the size of the crack; as the latter increases, the additives are also increased up to a point where use of mineral additives can be permitted (generally cracks exceeding 2-3 mm).

In the case of prestressed tendon sheaths, the technique is much the same, whether the operation is performed during the maintenance stage or at the time of construction. In recent years good results have been obtained in both France and Italy with vacuum injection systems following prior evaluation of the cavities present. This technique requires a good air-tight sealing of the structure, by applying an external sealing coat to the zone involved. In using epoxy resins as filler material, either they should be used in combination with cement so as to obtain a mixture pH sufficient to protect the steel (and in such cases the viscosity is quite high), or care should be taken to ensure the perfect adherence to the prestressed tendon or wire, as otherwise there would be a break in the continuity of the adherence allowing water to circulate and corrode the steel elements. The most promising technique is one preventing the entry of water in the first place, by perfectly waterproofing the bridge deck and painting the side parts.

III.2.6 Maintenance of metal structures

Most of the maintenance operations on steel structures can be defined as "traditional", in the best sense of the term: that is, the maintenance techniques are well-known and their results are satisfactory. In the following, only two types of operation will, therefore, be dealt with:

a) repair of fatigue cracks which occur more frequently on structures of modern design;
b) painting of steel components because this is the most widely-used maintenance technique and because new methodologies are emerging.

Fatigue Crack Repair

Some countries (in particular the United Kingdom and the United States) are beginning to experience some problems with fatigue cracks in various parts of steel bridges, especially the deck elements of orthotropic decks. A number of methods of crack repair are being used, ranging from the interception of the tip of the crack by the drilling of holes, the blunting of the crack tip by hammering or peening, or even the re-melting of a cracked area by welding techniques. Great care must be exercised in the evaluation of the crack repair methods used on the main structural members of bridges, since this involves the safety of the structure.

Painting of Metal Components

In this field there is considerable uniformity as regards assessment techniques, methods of surface preparation, and materials and techniques for painting operations.

This is due to the fact that this is one of the "oldest" aspects of modern structural maintenance, with the first large metal structures dating from the early 1880s.

As opposed to other maintenance sectors, there are standardised criteria for determining the degree of deterioration reached and, consequently, the preferable moment for performing the interventions. These criteria are generally based on the percentage of structure surface which has become degraded. Many countries follow the Swedish standards SIS 18.51.11, which permit determination of the degree of corrosion by means of a photographic comparison. The "European Scale of Corrosion Degree", prepared by AFNOR in France, and the American scale reported in the ASTM standards are also used.

The methods applied in the various countries for cleaning surfaces are also similar. They depend upon the age of the structure and the degree of degradation, and may consist of the following: washing and sanding to varying degree (generally on more recent structures), burning with special oxyacetylene torches with a high percentage of oxygen, followed by mechanical or manual brushing or sanding. There are also standards for determining the degree of degradation reached in relation to the types of protection which should be applied to the surfaces. The standards in the various countries are the Swedish SIS 05.59.00. The length of the time interval between these operations and the application of the first paintcoat depends on the following factors;

- climate;
- relative humidity;
- presence of aggressive agents.

In the countries surveyed the time interval chosen is in the range of hours, inasmuch as the surfaces treated can oxidize very quickly.

Even in one and the same country, both traditional and more modern types of anticorrosion paints are used. Base coat passivators include both traditional coatings such as minium (red-lead) or zinc, and mixtures of complex salts such as zinc chromates and phosphates and lead silicochromates, which provide more effective protection. These passivators are combined with different types of binders such as alkyds, oleoresinous/phenolic, chlorinated rubber, chlorinated rubber/alkyd and epoxy resins.

Use of the various types of binders often depends upon economic considerations, which are not only related to the costs of the painting cycle operations, but also to the varying thicknesses necessary to obtain comparable corrosion protection (number of coats). More recent formulations such as epoxy resins require a more careful preparation, but provide good protection with thinner layers.

Given the function required of the base coat and its high passivating salts content, it is better applied by brush or roller, so as to obtain better impregnation of the substrate; this view is shared by many of the countries surveyed.

Airless spray application is also permitted, however, for the subsequent layers.

All layers over the base coat will generally have the same composition as the latter, but with certain variations in the passivating salts content, inasmuch as they are normally covered with a finishing coat.

The finishing coat, when this is employed, is generally of a different composition than the undercoats, since it must be resistant to external agents (infra-red and ultraviolet rays, thermal shock, etc.), and must prevent as long as possible the penetration of degenerative agents into the underlying layer(s). The best finishing coats (and also the most costly) are polyurethane-resin-based paints which are often used to cover epoxy-resin-based undercoats. Chlorinated rubber-acrylic resin finishes are also used over non-mofified chlorinated rubber coats.

Paints used for anti-corrosive cycles can be checked beforehand by submitting them to artificial ageing tests in the laboratory (by means of weather-o-meter equipment, for example), so as to verify their durability prior to application. Chemical/physical tests are performed subsequent to these cycles in order to evaluate losses in characteristics with respect to the original materials. This procedure makes it possible to obtain an economic evaluation of solutions which are both more specialised and more costly than traditional ones.

III.2.7 Foundations

This section deals exclusively with the maintenance of foundations in water (river beds, canals, etc.), since special maintenance techniques are not required for dry foundations. In fact, interventions on dry foundation are nearly always of a repair or adjustment nature, and cannot in any way be considered as maintenance measures. Foundations immersed in water, however, present a very different situation, insofar as they are constantly subject to the effects of water currents, scouring, impacts of objects transported by the waters and, in some countries, ice.

Many rivers exert serious erosive action, and great efforts are required to safeguard foundations. In particular, alluvial channels will often tend to vary their position, thus having negative erosive effects.

When the river transports abundant debris the channel can often become blocked, thus causing damage to the superstructures of bridges as well as to the foundations, especially during periods of high floods. Maintenance to remedy damage caused by currents is very important, even though other types of structural damage may be more readily apparent to the eye.

Since most watercourses undergo constant change it is normally impossible to prevent alterations in their morphology. The designer must thus take these unknowns into account and anticipate their effects, adopting even at the design stage certain practical countermeasures to compensate for negative alterations.

It is no simple task to devise effective solutions. Settling of foundations, localised scouring, bank erosion, degradation of the river bed, etc. are complex problems which admit of no easy *a priori* answers. It is necessary to consult specialists in the fields of geology, hydraulics and structural engineering before hazarding a solution to a serious maintenance problem, because in some cases certain remedies may actually result in further harm.

Review of Erosion Problems

The most dangerous situation arises when the pier or the footing and the direction of the current are oriented in different ways. The erosion of the pier and consequent settling can result in serious structural damage. The erosion can also reduce the stability of the protective cutwaters insofar as it removes material from the zone of greatest exposure to friction. In certain cases even footing structures based on layers of resistant material can lose part of their resistance.

Scouring can also be the result of a narrowing of the current in the zone where the bridge is located, the shape and pattern of the eddies, the material of which the bed is composed, the material making up the banks, as well as the size of the piers themselves.

Serious lack of alignment between bridge and current, often attributed to errors in design, can actually be the result of a change in the morphology of the watercourse, which renders the structure unsuitable for the new conditions.

These changes may occur during exceptionally high floods, when heavy debris transported at great speed can tear up considerable quantities of material from the river bed, thus significantly increasing its depth. Scouring is a process which depends upon time, but is considerably affected by the variable nature of the currents. Its effects are evident after particularly intense floods. Many complex factors act together to produce scouring of river beds and changes in alignment and depth. It is thus necessary to consult experts before attempting to deal with problems connected with scouring.

Control of Erosion Effects and Protective Measures

The methods used are well-known and standardised in almost all countries; nevertheless, it is useful to recall certain general concepts in order to give a comprehensive panorama of maintenance techniques. The discussion hereunder is based on the detailed review of the U.S. AASHTO Manual for Bridge Maintenance.

Erosion can be controlled using various types of protection: rock or stone slopes, grouted rocks, inter-connected protections, stream retards, concrete blocks of different geometric shapes (e.g. tetrapods), gabions filled with stones. The type of protection should be compatible with the position and the natural aspect of the banks, and should make use, where possible, of locally available materials. In selecting the materials, the critical factors are the speed of the current, the material of which the bed is composed at that point, and the direction of the current. Before deciding on the type of intervention, it is necessary to fully analyse the phenomenon and the problems posed.

During floods erosion can be controlled by emergency measures, such as piles of rock or sandbags. Once the emergency is past, permanent measures can be taken if needed.

Large stones placed at critical points may be sufficient to counteract minor erosion phenomena. In more serious cases it will be necessary to have recourse to sheet piling of sufficient depth to extend down to non-erodible soil or rock.

In other cases it may be necessary to take measures to maintain the current in its bed beneath the bridge. If the bed upstream is winding, or shows signs of changing its course, one effective but costly solution is to dig a pilot channel to straighten the course of the river and then block off the old course so as to force the flow into the new course. Changes in the course only last temporarily, however, and the winding conditions will soon develop again.

Sometimes a winding course can be corrected by straightening the banks with slopes of hard rock. When the banks are of sand or silt soils, normally fine grain, it is necessary to take steps to prevent erosion of the material underlying the rocks.

In the past, thin layers of wood were used or the banks were stabilised by planting vegetation, but this provides only temporary protection and may be difficult to implement. In recent years it has become common practice in Europe to use layers of non-woven fabric of synthetic fibre (generally polyester or polyamide) which, thanks to their structure and small inner voids, can serve to hold in place even very fine material, thus permitting passage of water but preventing erosion beneath the bed of the rocks. These non-woven fabrics, easy to install, are today used for many purposes in hydraulic and geo-technical works, such as filtering, distribution of loads, reinforcing of soils, etc.

One of the most common methods for protecting banks and breaking currents is the use of gabions, which consist of metal nets filled with stones or coarse gravel. In making the decision whether to apply this technique, the following aspects should be taken into consideration:

- the technique requires considerable manual labour;

- the gabions are difficult to lay at depths exceeding half a metre;
- there must be local availability of rocks of suitable size and chemico-physical resistance;
- the metal nets must be provided with special protection if used in the presence of chemically aggressive waters or discharges.

As mentioned above, it is also possible to construct dry-stone or concrete protections. All these measures may have their weak points in the zones where they commence, and especially where they terminate. In fact, it often happens that the bank protection works have a roughness factor much lower than that of the original banks, with the result that the water velocity is increased as the erosive action is greatly diminished along the now protected banks. This action, however, is exercised with much greater intensity on the bed of the river, deepening it and eventually resulting in the formation of a race. Moreover, the greater velocity can continue for a certain distance downstream, with all the unknowns which this implies.

It is thus necessary to pay particular attention to the point of transition between the protected zone and that left in the original state because of the need to avoid undermining the base of the protective system and to prevent adverse effects on the downstream course.

The protections might consist of the following:

- a current cut-off wall to prevent scour action from damaging the outlet zone;
- a protective layer, possibly of non-woven fabric, to prevent scouring of deep-lying material;
- energy-dissipating devices to reduce the water velocity below the erosion level;
- any combination of the preceding systems.

Another cause of scouring both upstream and downstream is the occurrence of eddies and turbulence.

Finally, when using these systems for the protection of river banks, it is recommended that particular attention be paid to any possible effects resulting in scouring of the river bed itself.

Maintenance and Repair of Foundations

As regards work effected directly on foundations in river beds, it is firstly necessary to stress the importance of careful design of the foundations, i.e. shape and position, with respect to the direction of the current and the level of the river bed, which will render these structures less prone to erosion. If the design is inadequate or the river behaves in an unforeseen manner, a number of more or less serious problems can arise. However, as this study is limited to the maintenance field, reference is made only to less serious consequences such as slight undermining, erosion of materials of foundations and piers, etc.

The measures aim at restoring damaged zones, but also serve a preventive purpose such as described above for regularising river banks. The difference between these measures and those effected on other parts of the bridge is due to the conditions under which they are carried out, i.e. in the presence of water, even though in most cases the work will be done in the low-water season. If the work cannot be done under dry conditions, it is necessary to provide for the usual arrangements for works in the presence of water and on sub-foundations, even to the extent of using cofferdams or deviation canals. It should be kept in mind, however, that it is preferable not to increase the size of the reinforced concrete bed of the foundation footing, as this could lead to an increase in the scour action. It is advisable to effect restorations in keeping with the original dimensions of the works and possibly substitute the eroded material in such a way as to restore the shape of the riverbed.

A very interesting technique used against corrosion from salt water is the encasement of slender piles, mainly of concrete but also of structural steel, in a fibre-glass reinforced plastic form. These forms, which must be at least 3 mm thick, are placed around the pile and sealed at the bottom with water-resistant epoxy bonding. They are then dewatered and filled with the specified cement mortar and eventually left in place after completion of these operations.

The State of Florida (United States) has issued specifications on the structural properties and the resistance to ageing of these forms and of the prescribed mortar (see Figure III.4). Other useful techniques for the repair of spalled foundations are used in Norway (Figure III.5); the epoxy floating method, which permits the application of an epoxy resin layer without working in the water, is noteworthy.

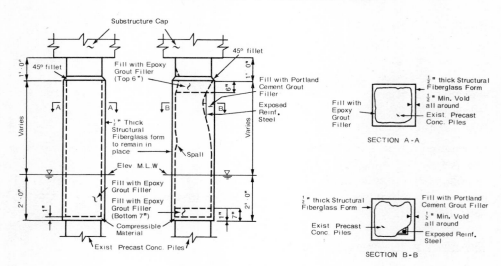

Figure III.4 EXAMPLE OF USE OF FIBERGLASS FORMS

Additional riprap or similar protection around the repaired piers represents a good solution or preventive measure, but this can lead to a shift of the scour zone downstream (cf. Figure III.6).

This can often be accepted, however, without particular harm. An alternate solution would be to install piles or rock protections upstream of the pier itself, as shown in Figure III.7.

The scouring of the material making up the pier is more easy to resolve, simply by lining the pier itself with material which is more resistant to the abrasive and disintegrating action of the water and the materials contained in it. The best systems are those which provide a stone cover, although there are others using coats of mortar and synthetic binder.

Figure III 5 **SPECIAL TECHNIQUES TO REPAIR SCOUR DAMAGES (NORWAY)**

1. UNDERWATER CONCRETING

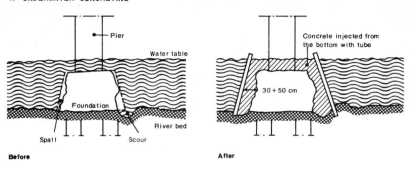

2. EPOXY GROUT REPAIRS

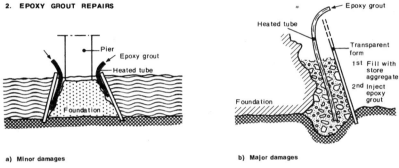

3. EPOXY FLOATING METHOD

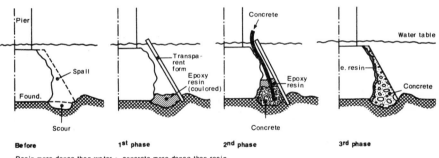

Resin more dense than water ; concrete more dense than resin.

Problems Caused by Sediments and Debris

Obstacles caused by sediment deposits can result in deviation of the current against a pier or abutment. This misalignment and flow concentration can cause a localised increase in current velocity, thus increasing its erosive action. Debris can accumulate against bridge members or trap other debris so as to form a partial dam. A sudden break can then cause overflowing of the banks, cut through the accesses to the bridge and damage the structure itself. To avoid such problems a number of maintenance measures are necessary which, however, are rarely undertaken due to their high costs. The following measures

Figure III.6 LOOSE RIPRAP

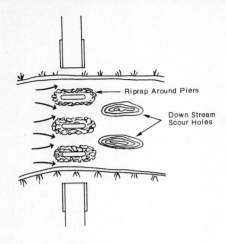

Figure III.7 SCOUR REDUCTION

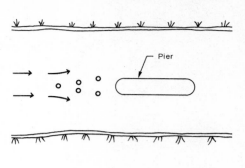

can be mentioned: manual clearing of the riverbed; elimination of any obstacles impeding the flow of water; removal of debris transported by the current; and, during floods, continuous removal of debris at the base of the piers.

<u>Problems Caused by Ice</u>

These problems are typical of northern countries, and frequently occur during thaws, when the ice breaks up and is transported by the current to narrow places where large blocks are formed. Jams can also be formed when the break-up of the ice is accompanied by a drop in the water level, such that the ice runs aground on the bottom and forms a dam. This latter case is the more dangerous because it totally blocks the water flow. The consequences over time may be serious both for the structure itself and adjacent zones, but there are no general rules for the control and prevention of this phenomenon. Each river has its own particular environmental and current characteristics. Necessary treatment in one point may be impossible elsewhere. The best prevention is to improve the bed in such a way that the ice cannot become blocked or run aground.

In some countries heat-absorbing materials such as carbon dust or chemical products are spread on the surface of the ice upstream, in such a way that the ice is weakened and does not block up under the bridges downstream. On the other hand, care must be taken to avoid reaching levels of pollution which might endanger aquatic life or drinking water supplies.

Before attempting to remove the ice, it is necessary to ascertain the possible damage which might occur downstream. Precautions need to be taken when using methods such as explosives for breaking up the ice. In the case of navigable rivers it is possible to break up the ice using tugs and ice-breakers.

There are also certain types of piers designed specifically for breaking up ice, which serve to prevent the formation of ice jams at the bridge itself. Sometimes it is possible to modify existing piers in this way.

III.3 SPECIAL TECHNIQUES FOR MAINTENANCE OF MASONRY BRIDGES AND TIMBER BRIDGES

For many centuries the only materials available for constructing spans of a certain size were stone, brick or combinations of the two for structures in compression or wood for structures in flexure. Results obtained with the use of these materials have been satisfactory and, indeed, timber bridges and especially masonry bridges can still be found in service today almost everywhere. It is thus appropriate to describe in this section at least the more important and widespread maintenance techniques for these types of bridges.

Timber bridges continue to be built in certain countries taking full advantage of new techniques to improve the performance of these materials. Construction of masonry bridges, however, has been practically abandoned, adding to maintenance problems in many countries insofar as there is an ever-growing lack of skilled workers with a knowledge of once-common masonry techniques.

III.3.1 Masonry bridges

In practice, masonry bridges include only arched structures or vaults in compression, whether they be constructed of stone, brick or a combination of the two. From the structural/functional standpoint and that of certain maintenance techniques, this category also includes arched structures of poured concrete.

The great durability of this type of structure is probably linked with the generous over-dimensioning which frequently characterises them, and which serves to protect them from fatigue phenomena as well as from vibrations which can aggravate normal degradation.

In spite of the presumed monolithic nature of such compressed structures, they are nevertheless quite delicate and must be supervised and maintained with care, as a defect at any point can have effects on the whole structure.

Typical defects of masonry bridges

In order to provide proper maintenance for these structures one must thoroughly understand them; unfortunately the technical details of this type of bridge and the relevant documents are often lacking. This problem can be resolved by carrying out careful surveys to determine the exact geometry and dimensions of the structure. Such surveys should also include an investigation of the nature and state of the component materials, since this information is indispensable if one wishes to evaluate the conditions of the structure. A point to be kept in mind, however, is the influence of the weight of the structure itself in determining the pressure curves of a vault.

Inspections of deficiencies should centre on the typical problems affecting masonry structures, which can be summarised as follows:

- problems arising from the materials and their usage;
- problems arising from poor construction arrangements;
- problems arising from changed operating conditions.

The first of these frequently occurs because the materials used in these structures are little understood and seldom utilised. Typical defects are those linked to the relative frost point and porosity of the stone.

A number of experts are of the opinion that degradation of the stone begins with the penetration of aggressive water or even rain water (i.e. deionised) into the voids of the material, with consequent dissolving and migration of the more soluble components to the outer surface; these crystallise on the outside in the form of scales and films. The presence of the latter indicates the initial stages of the process of degradation of the stone. The materials also become dangerously frost-susceptible. Other factors with harmful effects on stone include such things as mortising in the wrong direction, handling bulk materials without proper care and unsuitable dressing of the stones.

Certain problems also arise in the use of bricks: although these should be relatively more inert chemically, poor firing or poor quality clay can render them rather vulnerable.

Finally, the binders must be considered: mortars which have too high a cement content expand initially and may give rise to shrinkage problems at a later stage, whereas supports constructed with concretes which are too lean may develop deep cracks, such that they cease to be monolithic, and abnormal stress distributions may result. Moreover, all hydraulic binders are subject to attack by selenitic water, whether sea water or containing industrial residues.

Problems may also be caused by inappropriate treatments, such as the use of copings consisting of non-durable materials, or the incorporation of sidewalks enlarging the carriageway, without taking into consideration the lateral load distribution, and which may, thus, give rise to swellings and crumbling of the side walls. The most typical of such mistakes, however, is the restoration and relining of the intrados of the arch without consideration of the efficacy of the drainage system; this often has the effect of accumulating moisture within the supports and the structure of the vault. All these defects may result in constraints if occurring on a large scale between the various parts of the structure, longitudinal cracks due to overloads or thermal effects, transverse or oblique cracks and sinking of the piers; the latter will become apparent from geometric irregularities in the structure, often visible to the naked eye, and often resulting from loads passing on the road beneath rather than from loading on the structure itself. One must also not overlook possible poor hygroscopic conditions in the access embankments to the structure, which often result in yielding or swelling of the wing walls.

In inspecting masonry bridges it is naturally necessary to take all of these problems into account and to keep careful watch over the development of any defects discovered, paying particular attention to cracks and alignment alterations in the arch support zones.

Careful attention must be given to the efficiency of the drainage systems, for both surface and internal water; the latter, in particular, can cause considerable degradation and losses of the fine aggregates.

Finally, if the vaults are equipped with rudimentary plastic hinges obtained by the insertion of lead sheets, regular inspections should include as thorough a check as possible of the efficiency of these devices.

Maintenance operations

Before commencing any maintenance operation it is important to be certain that drainage is adequate and, if this is not the case, to install such systems or replace ineffective ones.

A distinction can be made between ordinary and special maintenance: the former includes simple operations such as cleaning of drainpipes, filling small cracks and replacing individual bricks or stones which may have fallen. The mortar used in such operations should contain less than 500 kg/m^3 of cement.

The following operations fall into the category of special maintenance.

a) Filling of cavities or cracks by injection

It is advisable to limit this operation to cracks no larger than 5 mm. Larger ones should be dealt with by means of partial replacement of the areas concerned. The materials used for these interventions are the same as those used for injections in concrete structures as described above.

Masonry can also be injected with cement products; in such cases care should be taken that the grouts have a low level of sedimentation and are non-shrinking. This can be ensured by means of the same techniques as those used for filling post tensioning ducts (high-speed mixers to prepare the grouts, which should be of the colloidal type to permit low water/cement ratios).

Care should be taken to avoid obvious mistakes such as injecting and thus blocking the drainage systems; a rigid bond between the masonry and the support should also be avoided, insofar as this could prevent the arch from moving freely in response to changes in temperature.

It should be kept in mind that simply filling in cracks in the structure of an arch, without attempting to tackle the causes of the cracking, is not only useless, but may well be counter-productive. For example, the relative settlement of the piers due to overloads on the road beneath the structure can produce cracks in the crown; these remain active and, if plugged, will re-form until such time as the causes cease, the settlement reaches its limit, or, better, the pier is reinforced.

Cracks in the arch can also give rise to more extensive breakdown processes, not directly related to the original causes. These, for example, can result in movements between the stones or bricks which in turn lead to irregular distribution of pressures in the arch. In such cases it is necessary to have resort to the suitable maintenance interventions described below.

Finally, it should be remembered that the structure and functioning of the arch are such that when horizontal cracks develop in the crown, wedge-shaped in the direction of the thickness starting from the intrados, it can be presumed that a similar cracking process is taking place in the springing; this is identical in every way except that it starts from the extrados, and thus appears to be less serious and more circumscribed, but is of similar size inside the body of the vault. Without going into a discussion of the origin of such crack patterns, which may be due either to overloads or to thermal phenomena, it must be kept in mind that however one decides to fill them, such cracks in the springing will require much deeper injections than might seem necessary at first sight. In particular, there is a proportionately greater risk that the injection material will disperse into secondary cracks in the masonry.

b) <u>Improvement of drainage and water evacuation systems</u>

Copings of waterproof materials were not common in the past, and consequently many surviving structures not so equipped are characterised by substantial and undesirable internal circulation of water.

In other cases the longitudinal slopes and crossfalls on the extrados are not sufficient to drain off the rainwater. Depending upon circumstances, it is thus necessary to equip the structure with gutters and pipes which will carry off the water without discharging it on the piers or supports, or with efficient outlets which correspond with the drainage system, if such already exists. Where such systems are lacking, a series of collectors can be installed at the points most susceptible to penetration by water; they may also serve to collect water circulating within the structure, in the absence of more thorough measures.

These installations, aimed at providing the type of protection currently given to modern bridges and to prevent future deterioration, are included in the category of preventive maintenance, in line with the distinction made in the preceding chapters.

It is evident that the zone to be treated must be carefully selected, possibly on the basis of suitable tests; similar care should be taken in choosing the diameter of the drainpipes, which may possibly be slotted or perforated and protected with a covering of non-woven fabric so as to limit the transport and loss of fine aggregates.

Once installed, these systems will be subject to ordinary maintenance, which will consist mainly in cleaning out the elements when they become clogged, for example by blowing out with water under pressure.

c) Installation of steel cross ties

These serve mainly to prevent the breaking away and bulging of the side walls with respect to the structure of the arch itself, but also to prevent the spread of possible longitudinal cracks.

The typical cause of these disorders is the widening of the carriageway and the incorporation of lateral margins or sidewalks, which previously were loaded rather lightly. These defects are very seldom the result of errors in the original design.

The cross ties can be very useful, but may also lead to unforeseen redistribution of forces; they should be employed with care, in suitable numbers and be of limited diameter. Close attention must be given in boring the holes for the ties: the methods will vary according to whether it is necessary to bore through the supports or the arch structure itself. The tie rods are protected with sealed sheaths; the bolts and anchor plates, which must be of sufficient size to prevent the concentration of stresses on the masonry facings, are provided with anti-corrosive protections such as described in Section III.2.6.

d) Installation of longitudinal ties

These serve mainly to increase the carrying capacity of the arch or to remedy a defect of the same.

A typical case in which they are employed is when a system of cracks develops in the crown and in the springings foreshadowing a yielding under load. They can also be utilised, however, in cases where the intermediate piers are out-of-plumb due to excess loading, but sometimes without the appearance of cracking in the crown. In this case the cracks in the springings will be longer and more extensive than usual, the breaks in the springing opposite the out-of-plumb pier being especially so.

It will be advisable to combine the installation of the tie rods with injections into the masonry. In general, the injections are performed first, and then the tie rods are put under tension and the mortar allowed to set. The filling material for the sheaths may be the same as that used for the injections.

e) Banding of piers

Maintenance work may not only concern the arch itself, but also the intermediate piers, and therefore the more common operation of banding is discussed hereunder. This procedure is especially effective in the rather frequently occurring case of piers consisting of dry masonry with mediocre filling with poor quality binders or when loose gravel is employed.

The banding must be installed in such a way that the resulting stresses on the pier are not dangerously concentrated at only a few points.

A sufficient number of hoops must thus be employed and these should not act directly on the pier, but on vertical elements installed at the corners of the pilaster, if this be polygonal, or regularly around the facing, if it be circular. To improve the strength of the banded pier, reinforcing injections may also be carried out.

It should be remembered, however, that if the pier has already begun to crumble at the base (a common effect of excess loading), producing typical convex-shaped fragments, it is necessary to carry out a major restoration operation, better classified as a repair rather than as maintenance.

f) Waterproofing of the extrados

In cases of partial or total restoration of arches it is possible to replace or repair a layer of waterproofing on the extrados. One should not neglect minor operations, however, such as injections of the masonry and proper drainage arrangements which can ensure the success of the operation.

The waterproofing layer should closely follow the profile of the arch, and be laid over a well-smoothed support layer. Capping layers of cement mortar are not recommended insofar as they may tear the waterproofing material.

The most suitable materials to employ, such as mastic asphalt and bituminous sheeting, are dealt with in Section 2.1 of the present chapter.

g) Repointing and restoration of masonry

Whether one is dealing with stone, masonry or brickwork, this operation is a rather delicate one, but still represents the most suitable way to restore the original appearance of the structure.

When restoring masonry, one must have a very clear idea of the lines of action of the loads, and one should always remember that a limited surface deterioration is often only the outward sign of a deeper problem. Appropriate shorings and supports must then be installed and, should the crack pattern be extensive, the shorings should be installed along the whole arch. In less serious cases some localised shorings can be applied, dimensioned and arranged in such a way as to take the forces along their axes and avoid buckling.

Similar shoring systems can be used also in the case of defects in the masonry of the piers. In either case the props are fastened to the masonry using wooden separators or wedges. In certain cases the vertical elements with masonry cracks may first be compressed between two well-connected and fixed sets of planking, but this constitutes only a provisional operation.

It is sometimes not necessary to replace the stones or bricks themselves, but only to replace the mortar bedding. In such cases one starts by carefully chiselling away all the old binder or by washing it out with high-pressure jets of water, possibly removing temporarily certain of the masonry elements. The resulting cavities should be cleaned, smoothed and dampened so as to receive the fresh mortar, which must be carefully inserted and compacted; in this case too it is preferable to use cement mortars with high cement ratios (at least 500 kg/m^3). Great care should be exercised to ensure gradual removing of formwork.

h) Protection of masonry

A very important aspect of masonry maintenance is the protective treatment of the stone materials.

As has been pointed out, the deterioration is very often related to their porosity, which exposes them to attack by atmospheric agents. Products for the treatment of stone, especially marl, sandstone and limestone, have been available for some time now, and consist of hardening materials, i.e. aqueous solutions of fluosilicates. This work must be done with care and above all using methods which ensure the penetration of the material deep into the stone, such as by immersion or prolonged pouring. One should avoid painting or spraying treatments, which often provide only a thin film which easily flakes off, leaving the stone exposed.

One of the most effective methods is deep-soaking with magnesium fluosilicates, which are not affected by rainwater. A small tube is inserted into each stone block so that it reaches roughly to the centre point; a vacuum pump is connected and allowed to run for a certain time so as to open up the pores and create a partial vacuum. Subsequently, the rock is sprayed with fluosilicate for sufficient time to permeate the individual stone block.

This method has been used with success in the past, but today it is also possible to utilise synthetic materials in such treatments.

This type of impregnation also serves to preserve the original appearance of the masonry which is often important for aesthetic reasons.

III.3.2 Timber bridges

The subject of wooden structures is dealt with separately from the others insofar as the related maintenance problems, though similar to those described in the preceding pages, have certain characteristics specific to the nature of the construction material. During the present century, in the majority of countries, wood is used less, for instance, than steel and cement concrete, which have become the traditional materials.

For a long time timber bridges were widely used along with stone arch and masonry bridges because, up until the beginning of the 19th century, wood constituted the only commonly used construction material able to support flexure. Even long after the advent of steel and reinforced concrete, timber bridges continued to be built, especially on secondary roads, because they offered the advantage of low construction costs, at least in those countries rich in wood supplies, and could be easily substituted. The most commonly used types of wood were oak, especially English oak, pitch pine, pine, larch and fir.

Agents affecting durability

The durability of wood, although not particularly good in comparison with other construction materials, can be considerably improved with suitable treatments, though it must be pointed out that these treatments may have harmful secondary effects on the mechanical properties of the material.

Maintenance measures for wooden bridges can be grouped in two main categories; either protective treatment against aggressive agents, or substitution of components which have been irreparably damaged.

Attempts to reinforce wooden structures with components of other materials, such as steel struts and tie rods, have normally met with limited success, since these latter, to be used economically, must transmit excessively strong tensions to their respective points of junction, which in turn become weak points.

The choice of protective treatment demands that, first of all, the causes of the deterioration should be determined. Some of these causes, such as atmospheric agents, are common to other construction materials, but their specific effects on wood may vary. Others, such as parasites, only constitute a problem in the case of wooden structures. Parasites in fact represent one of the greatest dangers to wood, particularly because of the fact that their action may also combine with that of other degrading agents such as salt water.

Although parasite damage can be considerably reduced and controlled through suitable preventive treatments, the structural parts which have already suffered severe attack generally have to be replaced, since even disinfestation cannot guarantee perfect results. Parasite attacks are particularly harmful in the presence of moisture, and thus constitute a grave danger to the structural members of bridges. Among the most harmful of these are insect attacks, particularly insidious because the wood may contain infestations even from the time when it was still at the plant stage. Insects may work in different ways, but the most serious are those which form colonies and spread with great rapidity. The fact that wood is transported from one place to another favours the spreading, cross-breeding and hence the strengthening of the many different species of parasites. The fight against these is thus a difficult one, and in many countries has necessitated the establishment of special agencies just for this purpose.

a) <u>Termites</u>

These are the most insidious type of parasite because the damage they cause lies concealed within the wood, leaving little or no superficial traces. Their presence can be spotted either from the crumbling or weakening of structural members or from the evidence of mud tunnels running from the soil to the wooden member.

These pests are considered so dangerous as to be the subject of special research on the part of the above-mentioned agencies. In Italy, for example, when structures are found to be infested with termites, this fact must be reported to the Experimental Institute for Agrarian Zoology in Florence, which in turn provides advice and assistance. In general this entails the discovery of the ways in which the worker termites managed to reach the wooden member. Once this has been done, then arrangements are made for the drainage and protection by means of barriers of concrete impregnated with certain products. Intensive disinfestation operations then follow, aimed at destruction of the nests, including those in the surrounding undergrowth and in nearby areas rich in wood residues, such as stumps and soil traversed by old roots. Obviously it is necessary to disinfest all the threatened structures, even if the latter will ultimately have to be replaced.

b) <u>Wood beetles</u>

These can be spotted by the holes in the wood surfaces and sometimes by traces of wood dust in the vicinity of the structure.

c) <u>Wood ants</u>

These leave piles of wood dust and can often be easily seen in the vicinity (all of the above insects attack every part of the bridge, but especially the piers).

d) <u>Marine organisms</u>

These include wood worms and limnoriae. Both of these act mainly at the water line of the piers, but generally have need of pre-existing holes in order to penetrate deeply into the wood. For this reason, it is advisable to provide metal or concrete protective guards at the base of the piers. Even unplugged bolt-holes can provide points of entry. The damage caused by these parasites may be such as to require the replacement of the piers, but suitable preventive treatments can almost always succeed in halting the infestation. When this does not succeed, the cause is usually to be sought in the lack of care taken in coating the more vulnerable points such as the ends of the braces, where the cross-section of the wood fibres is left exposed.

One of the most widely used means of guarding against the attack of animal parasites on wood structures is the impregnation of holes and pores with cyanide, keeping in mind that besides killing off the animals, it is also necessary to sterilise the eggs. Following this, there is the problem of covering up the holes, an operation which was traditionally performed with wax. Today this uneconomical and clumsy practice has been abandoned in favour of coatings with suitable sealant paints; the best are those based on tar substances containing disinfestants to prevent subsequent penetration and attacks.

e) **Fungi**

Fungi constitute another class of common aggressive agents, especially in damp conditions, where they are also accompanied by mould. The deterioration can be prevented only by careful preventive treatments. Since fungi and mould prosper in damp conditions, the points where they first become manifest are in the spaces between the planks which do not fit together perfectly and in the insets of joints and bolt holes (where they facilitate corrosion) and in the support zones.

Mould growth can be especially hazardous on the sidewalks where slippery patches can develop, constituting a danger to pedestrians.

Since moisture highly favours the growth of fungi, it is obvious that the formation of puddles and damp zones should be avoided, and great care should be taken to arrange for the disposal of rainwater.

The presence of fungi is revealed by dark-coloured spots and excessive porosity of the surfaces involved, which become soft and spongy, and even crumbly in the advanced states of deterioration.

Weathering

This is evidenced by superficial swellings or even by large cracks which might run completely through the member.

Wood used in the open air is highly affected by temperature and moisture variations. Exposure to the atmosphere results mainly in variations in volume, which are accentuated by variations in the wetting-drying cycle. Given the poor thermal conductivity of wood, the structures tend to contract during hot weather (when the drying out exerts a stronger effect than the heating of the wood), and expand in cold, wet weather (when the dampening-swelling effect prevails over that of the cooling). This phenomenon may be evidenced in all parts of the structure, resulting in the loosening of bolts and joints, but has particularly severe effects on the main beams. In this latter case the variations in volume, together with the tensile stresses, result in a progressive slackening of the stretched fibres to the extent of opening cracks, which often can be simply the continuation and deepening of small secondary breaks which may have developed during the seasoning period.

The flexural members thus become extremely unstable and particularly sensitive to vibrations; their deflection increases dangerously. This state of affairs is often indicated by the appearance of small cracks in the asphalt paving layer, which form an almost continuous network over the whole road surface, termed "alligator" cracking. Besides close visual examination, a good way to ascertain this type of deterioration is by observation from underneath when the bridge is being traversed by heavy vehicles.

It should be emphasised that atmospheric agents act not only on the wood but also, more importantly, on the metal connections and bolts which are subject to rusting. The only remedy here is the exclusive use of galvanised hardware.

The paint coatings on the wooden members may also undergo weathering. Such coatings on wood are not as important as those on metal, as they do not serve a protective function, but they serve as useful visual guides to drivers, especially on the guardrails. Paints should always be applied only after adequate preservative treatments, and only when the latter coats are completely dry.

The other causes of ordinary deterioration - apart from the effects of collisions, fires, or overloading - are mechanical abrasions and friction, which obviously only affect the planking of the sidewalks and/or the roadway where these are left uncovered. These result from normal use of the bridge, but may be aggravated by the presence of dirt, sand, etc. on the surface. When worn planking is being replaced with new, the latter members should be arranged so that their grain parallels the traffic direction, ensuring that they are resting perfectly on their crossbeams.

Maintenance operations

The various maintenance operations are described and grouped below, according to the various parts of the bridge which are concerned. This classification follows the most rational way of dealing with timber bridges, whose structure is easily divided in its main elements.

A. Maintenance measures for deck slabs and surface layers

a) The traffic runs directly on the wooden deck

Timber bridges which are situated on local roads or on other roads with minor traffic as well as bridges for pedestrian and cyclist traffic are usually equipped with a pressure-treated wood deck without any surface materials. Two types of decks are commonly used, namely:

1. Nailed laminated deck (planks directly nailed to each other or with intermediate ventilating spaces);
2. Longitudinally and/or diagonally nailed layers of planks on crossbeams or stringers.

The use of these types of timber deck is justified if there is only light traffic on the bridge or if the volume and axle loads of the vehicular traffic are moderate. However, if the vehicular traffic is considerable or heavy, serious maintenance problems often arise.

A maintenance problem for decks of the first type (laminated) is:

- Excessive wear of planks. Steel cover plates can then be used over the wheel tracks. This measure can be preventive - taken in advance - or a repair measure used when excessive wear of the wheel tracks has been observed.

In Finland steel cover plates of 1,200 x 4,000 x 5 fastened to the plank deck by bolts of 10 mm. diameter are normally used. To muffle the excessive noise a bituminous felt is placed between the steel plate and the wood deck. The steel cover plates must be grooved to prevent skidding.

Usual maintenance problems in the case of the second deck type (longitudinally and/or diagonally nailed plank layers) are the following:

- The nails work themselves loose with the continuous traffic movement. The extruding nails are not only hazardous for tyres, but also affect the load bearing capacity and service-life of the bridge as well as creating extensive maintenance work in renailing of loosened planks.
- The routine maintenance generally involves replacement of decayed or worn-out deck planks and crossbeams.
- To prevent excessive wear, steel cover plates or alternatively elastomer surfacings may be used or the decks may be surfaced with asphalt layers.

b) <u>Wearing surfaces on wooden decks</u>

On roads with heavy traffic the wooden decks must usually be furnished with some kind of a surface coating to prevent excess wear of decks. The following surfacings are most frequent:

- bituminous surfacing;
- concrete surface layer acting separately or as a composite system with the wooden deck;
- special coatings.

The following comments on the maintenance of the above coating materials can be made:

1. Bituminous surfacing on wooden decks
   The most frequent maintenance problems are those caused by dynamic action of traffic loads. Due to vibration of the wooden deck the bituminous layer tends to loosen from the bottom and this leads to cracking, spalling and slipping. To prevent such damage various precautions have been used, such as:
   - coating the wooden deck with a bituminous adhesive, e.g. cut-back bitumen before laying the asphalt;
   - using a steel net in the asphalt layer. The net is fixed to the wooden deck with nails or hooks.

   Maintenance work includes the repair of cracks and spalls in the surfacing using adequate hot bituminous mixes or suitable sealing compounds. This prevents the water and debris from penetrating through the defects to the wooden deck, which would result in rot.

2. Concrete layers on wooden decks
   In the United States and Canada an alternative method of construction is to use a concrete layer on the wooden deck to serve as a wearing surface or as a structural part of a deck system. This is called the concrete-wood composite system.
   Maintenance measures for concrete surfaces are fundamentally the same as for concrete pavements generally, like crack injecting with epoxy-resins, patching of spalled concrete, etc.
   Possible preventive measures can include impregnating concrete surfaces with suitable agents to prevent corrosion by de-icing salts.

3. Special coatings on wood decks
   Today there are special products for preventing excessive wear of wooden decks. These products are elastomer or polyurethane-based materials. However, information about them is scarce.

c) <u>Strengthening and rehabilitating separated laminated decks by prestressing</u>

In the laminated system, adjacent deck planks are nailed to each other. However, it is possible that under repeated heavy loading the nails bend or crush the surrounding wood. This condition progresses so that the laminates separate allowing the ingress of water and debris. The resulting situation is that the laminates no longer act together to distribute the load and local failures occur under overloads (Figure III.8). But this condition also leads to an increased rate of deterioration which will eventually lead to failure regardless of overloading. In the Province of Ontario, Canada, the service life and load bearing capacity of old wooden bridges has been increased significantly by applying transverse prestressing to longitudinally laminated bridge decks. The post-tensioning

Figure III.8 **SEPARATION OF LAMINATED TIMBER DECKING DUE TO LOCAL OVERLOADS (ref. H5)**

Figure III 9 **TRANSVERSE PRESTRESSING OF LONGITUDINALLY LAMINATED BRIDGE DECKS (ref. H5)**

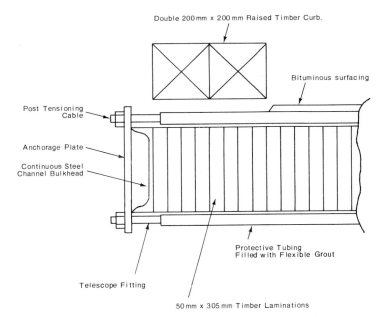

was achieved by using the same high strength bars as used in prestressed concrete construction (Figure III.9). The post-tensioning offers a long-term solution for minimising the interface movement between laminates. This idea, originally only intended for rehabilitation, is now being implemented into new designs.

B. Maintenance measures for truss structures

Normally truss structures do not create maintenance problems because the timber in these structures should be absolutely sound. If this is not the case the whole superstructure or bridge must usually be replaced or removed. If there is reason to suspect the quality of timber material, bored samples of the wood should be taken to detect possible decay or damage caused by fungi, termites, marine borers, shipworms, etc.

The following maintenance measures may be needed:

- Retightening of loosened bolts or other connection members;
- Injecting epoxy in major cracks in glulam beams;
- Varnishing exposed surfaces of glulam beams to protect from excess sunshine;
- Inspecting the tightness of bearings in continuous structures. If there is free space between the superstructure and the substructure, for example on intermediate pile-bents, the risk of dynamic traffic loads driving the piles even deeper is great. Suitable measures should be taken such as using wedges;
- Where bolted connections are used conditions such as cracking can occur. One method of repair, tested by the United States, is epoxy-repaired bolted connections. However, if defects are severe, the member must be replaced
- Timber stringer replacement.

C. Maintenance measures on substructures

The following maintenance operations may be needed for substructures:

- Protecting piles from excessive wear caused by moving ice and floating timber, etc. The protection usually consists of a plank wall, steel plates or angles;
- Retightening of bolts in bracings of pile-bents and abutments;
- Supporting tilting pile-bents or abutments by props;
- Preventing the soil behind abutments (frontwalls and wings) from leaking between planks and logs. If no measures are taken the road surface might sink behind the abutment and thus jeopardise traffic safety;
- Replacing rotting or otherwise defective piles and caps.
- Often timber is used in dolphins or fenders which are used to protect the substructure against collisions. In such cases abrasion, weathering and decay are problems.

III.4 AUXILIARY EQUIPMENT AND TECHNIQUES

In all countries it was noted that very often the main problem when repairing bridges is precisely that of reaching the damaged locations. In fact, the erection of traditional scaffolding, especially in cases of decks located high above ground level, could constitute the major part of the cost of the repair operation, and would thus be unacceptable from the economic standpoint. Operating and maintenance experience has shown the advantages of considering the access problem at the bridge design stage (see Chapter IV).

Sometimes on the occasion of essential maintenance works, access systems are installed at the more vulnerable points, even on older structures which were not originally so equipped. Sometimes it is advisable that access be extended to all parts of the bridge, even for locations far from the bearings, by means of passages formed in the structure itself.

Ensuring the practicability of reaching all parts of the structure, however, does not in itself resolve the problem of the working equipment, since the points in need of repair such as the intrados of the beams, are not accessible to possible inspection. The solution to this lies in the use of mobile scaffolds able to move along the structure and permit small work crews to reach all points of the intrados with their equipment, remaining in constant telephone contact with the personnel responsible for moving the scaffold. An example of this mobile scaffold is shown in the attached drawing (Figure III.10).

There are also other similar types of equipment of smaller size which lack self-propulsion systems, but which can easily be mounted on small trucks.

The device shown in Figure III.11 can reach the intrados of a bridge to a depth of 10 metres, and has the advantage of a moving arm with a 90° articulated end-piece which makes it possible to work in close contact with the intrados even when the main beams of the deck are quite high.

Figure III.10 **EXAMPLE OF MOBILE SCAFFOLD**

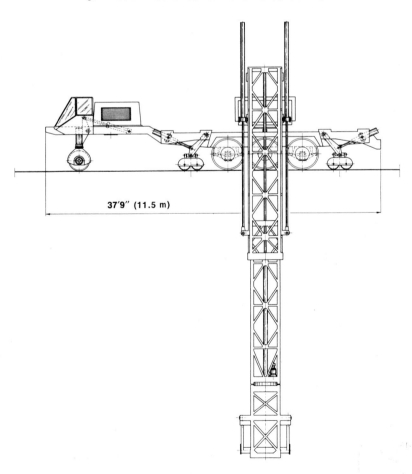

Figure III.11 **WORKING RANGE OF A BRIDGE LIFT**

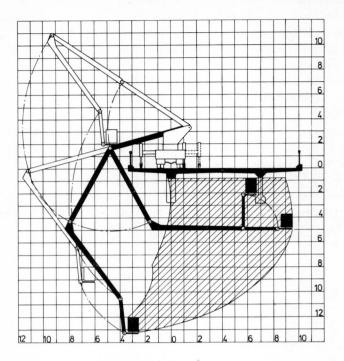

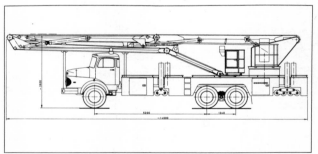

III.5 CONCLUSIONS

Examining the experience of the countries participating in this study, it is possible to draw some preliminary conclusions which will be analysed in more detail in Chapter IV. These concern:

- the importance and need to encourage preventive maintenance, overcoming the economic limitations often due to shared management responsibility of the agencies involved;
- the maintenance of the large number of old bridges, constructed with techniques no longer in use, but which have proved to be highly resistant to fatigue, probably because of initial overdimensioning;
- the great importance of maintaining waterproofing, expansion joints and supports when preserving structures, as this will help to prevent growth of defects;

- the more frequent use of synthetic materials, given their fast-setting and good physical characteristics, as opposed to the lack of information on their long-term reliability and the absence up to now of standards regarding their use;
- the need to train specialists in maintenance science (terotechnology) which is now done in certain countries using teams comprising various experts in a number of fields (engineers, geologists, chemists, physicists, data processors, generally co-ordinated by engineers).

BIBLIOGRAPHY

A. GENERAL BIBLIOGRAPHY

1. OECD, Road Research. Bridge Inspection, Paris, 1976.
   Economic and technical aspects of bridge structure management; identification of the various structural elements most in need of maintenance; economic and finance problems; staff training and tasks; utilisation of results by means of computerised systems; systems of access to repair points; testing methods presently used with evaluation of their utility; proposed classification of inspections according to their purposes and frequencies.

2. MINISTERE DE L'EQUIPEMENT. Service d'études techniques des routes et autoroutes, Laboratoire central des ponts et chaussées. Défauts apparents des ouvrages d'art en béton, France.
   Classification of all of the more easily found defects, broken down according to their origin, the probability of their worsening, the possibility of their having negative effects on the structural behaviour of the bridge; in the case of each defect a photographic documentation, an index of seriousness, the cause of its appearance and its probable evolution are provided.

3. GENTILINI B. and AGHILONE G. Problemi della costruzione e dell'esercizio di opere d'arte per le autostrade in terreni montagnosi. Rivista Autostrade, Italy, December 1976.
   Evolution of the concept of bridge structure construction economy, with abandonment of savings on materials in favour of accurate fundamental choices which permit rational use of mechnical means; prestressed reinforced concrete construction systems utilised by Autostrade Company; consequent classification of the bridges managed by the Company, on the basis of their construction systems, the materials used and the period of construction; maintenance problems encountered.

4. U.S. DEPARTMENT OF TRANSPORTATION/FEDERAL HIGHWAY ADMINISTRATION. Bridge Inspector's Training Manual.
   Text recommended for the training of structure inspection personnel. The first part contains the nomenclature for the various bridge elements, the classification of bridges on the basis of their types and of the materials used, as well as basic principles of construction science. The second part consists of a detailed examination of the types of damage characteristic of the various structure categories, with useful suggestions on how to pinpoint these, as well as how to organise the information obtained in a complete and exhaustive file.

5. FLORIDA DEPARTMENT OF TRANSPORTATION, State Maintenance Office, United States. "Manual for Bridge Maintenance Planning and Repair Methods". Volume I - Repair Methods.
   Manual containing the classification of the main problems encountered on Florida bridges, with detailed descriptions and illustrations of the various stages. All the methods described are easy to employ and require no special means or raw materials, and thus can be utilised also by outlying maintenance posts.

6. AMERICAN ASSOCIATION OF STATE HIGHWAY AND TRANSPORTATION OFFICIALS. AASHTO Manual for Bridge Maintenance, 1976.
   Exhaustive classification of the damages typical of each element of a bridge structure, with indication of the systems for identifying these, the probable causes and the corrective intervention measures.

7. ASSOCIATION FRANCAISE DES PONTS ET CHARPENTES. Journées nationales, Theme III, Conception et durabilité (rapport), Paris, 1978.

B. BIBLIOGRAPHY ON PAVEMENTS AND WATERPROOFING

1. DE BAKER C. Catalogue des dégradations des revêtements hydrocarbonés d'ouvrages d'art, Centre de Recherches Routières, Brussels.
   Classification, causes, possible evolution and remedies for main types of pavement degradation; includes a precise definition and photographic documentation of each type of defect.

2. OECD, Road Research. Waterproofing of concrete bridge decks, Paris, 1972.

3. CAMOMILLA G. and PERONI G. Impermeabilizzazioni di ponti e viadotti con guaine bituminose armate. Rivista Autostrade, July-August 1976.

4. RIO A. and CERNIA E.M. Materiali compositi di calcestruzzo cementizio e polimeri organici. Industria Italiana del cemento, magazine, year XLV, May 1975, and Macromolecular Reviews, Vol. 9,00-00 (1974).
   Experiences in and methodologies for obtaining extremely resistant and durable concrete by means of impregnation with thermoplastic monomers (PIC). This is a general treatment, i.e. it is not directed particularly at the solution of road problems; it does however provide useful indications for application in the maintenance field.

5. BENSE P. Application d'un enduit superficiel comme technique d'entretien d'un ouvrage d'art. Dalle béton - Dalle orthotrope. Bulltin de liaison des Laboratoires des Ponts et Chaussées, No. 70, Paris, March-April, 1974.

C. BIBLIOGRAPHY ON PROBLEMS WITH JOINTS

1. AGHILONE G., CAMOMILLA, G. and FREDIANI G. Misure dei movimenti delle opere d'arte in corrispondenza dei giunti di dilatazione. Rivista Autostrade, Italy, March 1975.
   Theoretical and experimental survey of stresses produced on joints by vibration movements caused by heavy traffic; consequent formulation of indicative criteria to be adopted in the design of the joint and of the structure.

2. BOTTURA N., BOCCATO F. and CAMOMILLA G. I Giunti di dilatazione nelle opere d'arte. Italian report to the Motorway Day of Seville, published in Rivista Autostrade, Italy, December 1975.
   Evolution of the problem of the joint from the first Italian experiences until today; classification of the various types of joint on the basis of the degree of expansion permitted; breakdown of joints on the basis of their various functions; trends emerging from motorway management experiences and experiments carried out.

3. KOSTER W. Fahrbahnübergange an Brucken und Betonbehnen, Eyrolles, Paris.

4. SETRA-LCPC - Environnement des appareils d'appui en élastomère fretté. Paris, October, 1978.

D. BIBLIOGRAPHY ON PROBLEMS OF WATER DISPOSAL ON BRIDGE STRUCTURES: PARAPETS AND GUARDRAILS

1. DEPARTEMENT FEDERAL DE L'INTERIEUR. Routes Nationales Suisses. Projets standards de ponts - Détails de construction, Switzerland.
   Collection of typical bridge projects considered from the standpoint of the main construction elements: codification of alternative solutions adopted in Switzerland, with illustration of details.

2. OECD, Road Research, Waterproofing of concrete bridge decks. Paris 1972.

E. BIBLIOGRAPHY ON RESTORATION OF CONCRETES

1. AUTOSTRADE COMPANY. Special Specifications for the painting and protection of concretes, Italy.

2. MINISTERE DE L'EQUIPEMENT ET DE L'AMENAGEMENT DU TERRITOIRE. Service d'Etudes Techniques des Routes et Autoroutes, Laboratoire Central des Ponts et Chaussées. Choix et application des produits de réparation des ouvrages en béton, France.
   Theoretical introduction to the problem of binders, with particular reference to organic ones; list and documentation of the main types of concrete degradation, with identification of both recommended and inadvisable products.

3. SIA - SWISS SOCIETY OF ENGINEERS AND ARCHITECTS. Technical Standard 162 for the calculation, construction and execution of simple, reinforced and prestressed concrete structures.

4. ROSSETTI A. and CAMOMILLA G. La degradazione del calcestruzzo nei manufatti autostradali: ripristini con malte speciali. L'Industria Italiana del Cemento, Italy, November 1978.
   Brief summary of the main degradations of concrete structures; identification of the causes of said degradations in the design of the works and in the quality of the concretes; formulation of rheoplastic mortars with improved characteristics; results of comparative experiments with rheoplastic mortars and epoxy resin mortars.

5. CAMOMILLA G. Le resine epossidiche negli interventi di manutenzione delle strutture autostradali. Atti del Convegno. Le resine epossidiche nell ingegneria strutturale. Collegio degli Ingegneri di Milano.

6. RIO A. and CERNIA E.M. Materiali compositi di calcestruzzo cementizio e polimeri organic (cf. above).

7. BERISSI R. and LAZZERI L. Expérimentation d'un nouveau procédé de réinjection des gaines des câbles de précontrainte. Bulletin de liaison des Laboratoires des Ponts et Chaussées, No. 82, Paris, March-April 1976.

F. BIBLIOGRAPHY ON PAINTING METAL STRUCTURES

1. MINISTERE DE L'EQUIPEMENT ET DE L'AMENAGEMENT DU TERRITOIRE, Laboratoire Central des Ponts et Chaussées. Guide de contrôle de chantier de peintures sur ouvrage métallique, France.
   Norms for the organisation of operations for painting of metal structures with regard to the examination of the works in question; precautions to be taken before painting, methods and controls for the cleaning and preparation of the surfaces; systems for carrying out the work and for checking its quality, final inspection.

2. DEPARTEMENT FEDERAL DE L'INTERIEUR, Routes Nationales Suisses. Traitement des surfaces des constructions en acier, Switzerland.

3. SWEDISH STANDARDS SIS 05.5900 and SIS 18.51.11.

4. AFNOR. European Scale of the Degree of Rusting for Antirust Paint, France.

5. ROADS AND WATERWAYS ADMINISTRATION. General Specifications for Bridge Construction, Finland.

6. AUTOSTRADE COMPANY. Special Specifications for the execution of anti-corrosive painting to protect steel structures, Italy.

G. BIBLIOGRAPHY ON PROBLEMS OF FOUNDATIONS ON FLUVIAL OR MARINE SITES

1. Bridge Inspector's Training Manual (cf. above).

2. Manual for Bridge Maintenance Planning and Repair Methods. Vol. I. Repair Methods (cf. above).

3. AASHTO Manual for Bridge Maintenance (cf. above).

4. MINISTRY OF PUBLIC WORKS. Compilazione dei progetti e norme esecutive per la sistemazione dei corsi d'acqua. Circular no. 6122 of 3rd July 1969, Italy.
   Indication of the purposes for regularising fluvial basins and techniques to be used; description of the preliminary research and studies, including the surveys and reliefs

to be taken, especially in relation to erosion and deposit problems; description of the protection structures, their technical and construction details, and of the criteria for the use of each category, also in view of the characteristics of the current involved; formulas and methods to be used to determine the main characteristics of the currents and for the dimensioning of the structural elements or to calculate the stresses to which they are subjected.

H. BIBLIOGRAPHY ON PROBLEMS OF WOODEN STRUCTURES

1. Bridge Inspector's Training Manual (cf. above).

2. Manual for Bridge Maintenance Planning and Repair Methods (cf. above).

3. AASHTO Manual for Bridge Maintenance (cf. above).

4. MASTRODICASA S. I Dissesti statici delle strutture edilizie.
   Complete treatment of all types of breakdowns, ranging from the theoretical setting up of the diagnostical problem to the description of the intervention technologies, subdivided according to materials and range of application.

5. A Development Program for Wood Highway Bridges. 79-SRR-7, Ontario, Canada.

6. DEPARTMENT OF TRANSPORTATION, FEDERAL HIGHWAY ADMINISTRATION. Standard Plans for Highway Bridges. Volume III, Timber Bridges, Washington D.C., 1979.

I. BIBLIOGRAPHY ON PROBLEMS OF MASONRY STRUCTURES

1. MASTRODICASA S. I Dissesti statici delle strutture edilizie, Hoepli, Milano.

2. MINISTERE DES TRANSPORTS. Surveillance et Entretien des Ouvrages d'Art, Ponts et Viaducs en Maçonnerie. France.

3. CENTRE DE RECHERCHES ET D'ETUDES OCEANOGRAPHIQUES. Colloque sur le mode d'action des produits et techniques de protection des pierres en oeuvre. La Rochelle, 1976.

4. PIET YH and CHAMPION M. La répartition de la huitième arche du pont Jacques Gabriel sur la Loire à Blois. Bulletin de liaison des Laboratoires des Ponts et Chaussées No. 101, May-June 1979, Revue générale des routes et des aérodromes No. 555, Paris, July 1979

Chapter IV

MAINTENANCE POLICY - PROPOSALS

IV.1 FRAMEWORK OF MANAGEMENT

IV.1.1 General conception of maintenance

Bridges are important, costly and sensitive elements - links - in the road systems. They are built to serve road traffic and traffic requirements that govern the roads do also govern the bridges. Bridges should be built, maintained and repaired to fulfil such requirements while methods and techniques need to be improved to the economic advantage of the community.

Bridge maintenance is part of the general road maintenance policy as well as part of a general bridge policy, the latter setting the standards from the first design phase, through construction and maintenance to the final demolition. The concept of bridge maintenance shall fit into both these general policies, and it can therefore hardly be expected that a concept of bridge maintenance can be formulated which would suit all bridge authorities in the OECD countries. However, although emphasis may be put on different issues, a great many of them are common.

Maintenance covers every operation aimed at maintaining a bridge in a state of serviceability. This is characterised by the bridge's load carrying capacity and by the level of comfort and functional security (both for users and those in its immediate vicinity).

IV.1.2 Management and principles of organisation

Assignment of responsibility between different authorities

Bridges are associated with a transport network of which they form part. The maintenance of bridges should be the responsibility of the authority (owner) usually responsible for maintaining the associated network.

Bridges and similar structures may often be common to two networks, e.g. where a road crosses over or under a railway line, a waterway or another road belonging to a different network. In such cases the maintenance should be carried out as far as possible by the authority of the overpassing network. In other cases the following rules should be observed:

a) The inspection and maintenance of any given bridge should be the responsibility of a single authority only, and work on different parts of the bridge should therefore not be divided between network authorities. The frequent practice of having one authority in charge of maintaining the actual structure and another of maintaining, e.g. the bridge's pavement should be severely discouraged.

b) When one of the authorities concerned is considerably better equipped than another to inspect and maintain a bridge, preference should be given to the former in carrying out these tasks. An example would be a bridge over a main road carrying a minor rural road belonging to a local authority with limited technical resources.

It should, however, be noted that the implementation of this measure can raise complex administrative or legal problems (such as the freedom of local authorities in the choice of technical assistance).

c) If a bridge requires particular technical competence of one of the two authorities concerned, it is advisable that this authority should undertake the inspection and maintenance of the whole bridge. Where a road is overpassed by a railway bridge, it is the railway authority which should be responsible for inspection and maintenance, because the bridge poses problems peculiar to the railways; in the case where a road passes over a railway the matter must be settled by agreement taking into account the means available to the authorities concerned (cf. b) above).

d) The authority responsible for inspection and maintenance should keep the other authority informed with regard to all work carried out on the bridge.

Finance of maintenance

When a road or railway involves building a bridge over an existing road or railway, total expenditure is generally borne by the authority requiring the new structure. In certain cases (e.g. bridges or underpasses where there was previously a level crossing) the two authorities may come to an agreement on the breakdown of expenditure.

Sharing of maintenance costs between authorities should however be avoided. It is a sound management principle that the technical responsibility is intimately linked with the financial responsibility, and in this context it is important to realise that the structure does not need to be maintained by the authority requiring the original construction. In other words, the organisation and management of bridge maintenance should be arranged in such a way that the maximum benefits are obtained from the available overall resources, overriding any particular or private interests of individual network authorities.

Organisation of maintenance

Routine maintenance should normally be carried out by teams working under the close supervision of the maintenance department staff.

Specialised maintenance cannot in most cases be performed directly by such personnel and indeed there is often little point in equipping these services for such tasks. As a general rule, specialised maintenance operations should be carried out, often under contract, by specialists with the appropriate equipment; these tasks should be identified and examined by the department in charge of maintenance, which should also ensure that they are properly carried out.

IV.2 MAINTENANCE POLICY

Maintenance policy includes three important considerations:

a) Safety considerations

The structural safety of bridges is an important issue although the influence of structural hazards of bridges on traffic safety is in general limited. However, the public's demand for a high threshold against structural failure must be considered, even if funds could be used more profitably on alternative actions for the improvement of traffic safety.

The general risk pattern must be analysed and evaluated, and a policy formulated that strikes a balance between the accepted residual risks and the efforts to confine such risks.

b) <u>Considerations relating to the flow of traffic</u>

Road bridges should serve the traffic to the same standards of accessibility, serviceability and comfort as the adjoining roads. Maintenance works should be done as far as possible without hindrance to traffic, and the bridge components should be designed with reasonable durability to minimise traffic restrictions during necessary repairs.

c) <u>Economic and technical considerations</u>

Bridges - like roads in general - are long-term investments. What is decided, done and planned today will reach deeply into the next century. Bridges must therefore be built and maintained with a high degree of foresight to avoid premature replacements that may prove expensive not only in the costs of reconstruction, but also in possible temporary closure of important traffic arteries.

Bridges deteriorate, and different parts of a bridge deteriorate differently. Bridge maintenance must apply a variety of specialised techniques to cope economically with the preservation of the colossal investments that the developed nations have poured into their infrastructures in form of bridges.

As these investments approach their peak with the completion of the motorway systems, the need for effective and economical maintenance becomes evident.

These three considerations must be seen in a global context - in a balance in which the three considerations may carry slightly different weight in different countries. To give prominence to, say, the economic considerations will offset this balance and may lead to adverse results. The order in which the considerations are mentioned is, however, not incidental. Safety does prevail over traffic, which again prevails over economic and technical considerations.

IV.2.1 <u>Elaboration of a maintenance policy</u>

The maintenance of a bridge starts on the day the contractor leaves the construction site, and it continues as long as the road that the bridge is built for exists and is in use, which might be indefinitely. The bridge might be rebuilt or replaced by a new bridge, and thus we enter into a life cycle of the bridge, that might or might not be dependent upon changes of the road it serves.

When a major repair is considered necessary - perhaps due to neglect of proper maintenance - it may prove advantageous to strengthen the bridge at the same time. However, it is not easy to distinguish between what is maintenance and repair, and what is strengthening.

Bridge management includes all operations needed to keep the bridge in good order till it is replaced or abandoned, namely:

  i) Conservation of the original facility.
 ii) Repair of damage due to lack of maintenance.
iii) Repair of accidental damage.
 iv) Widening or strengthening to provide an improved facility.
  v) Reconstruction or demolition.

The present report deals mainly with the points i), ii) and iii).

The background and reasons for widening and strengthening a bridge are related to alterations in regulations of traffic loads and in the classification of roads. The techniques applied in such cases relate to specialised maintenance technology and the special problems in point iv) should therefore not be entirely excluded when dealing with the technical problems of maintenance. Furthermore, the costs of v) must be taken into account in evaluating alternative maintenance strategies.

The considerations mentioned above

- safety
- traffic
- economy and technology

imply the following, i.e.

- satisfactory design and construction;
- effective bridge inspection including thorough documentation;
- evaluation techniques - including the recording and retrieval of cost data - enabling the bridge authority to act and decide when measures are appropriate;
- a suitable maintenance organisation qualified to implement the maintenance policy;
- training and research.

IV.2.2 Safety aspects

The safety of bridges is considered an important issue in bridge maintenance policy. To bring this issue into the perspective of the overall bridge management, the risks must be considered objectively.

The risks to human lives arising from negligence, malfunction or collapse of bridges are indeed very small compared with the general risk pattern of human behaviour.

There exists only a few shaky estimates of these risks. W. Bosshard cites CIRIA in the 1979 IABSE paper 5-9/79 "Structural Safety - A Matter of Decision and Control":

- The annual risk of death of any person in the United Kingdom due to collapse of a completed structure from any cause is estimated 1:7,000,000. This risk should be compared with the 1:10,000 risk of the same person to die from a traffic accident, the 5:10,000 risk of a construction worker to die at work in any given year, and the general mortality of the population, which is 1:1,000 at age 30 and 6:1,000 at age 60.

Naturally, the risks arising from collapse of bridges are only a fraction of the above-mentioned yearly 1:7,000,000 for all completed structures.

Risk levels thus vary greatly within our daily lives and their acceptance by the society is governed by the following factors:

- the degree of voluntariness or payment for the persons exposed to the risks;
- the necessity of acceptance of risks to avoid impediment or increase in costs of important public interests;
- the technical feasibility - within acceptable economic limits - to reduce the risks.

In this context there is no doubt that preservation of structural safety is an important issue in bridge maintenance, not only because it is technically feasible to safeguard the structures but also because it is economically sensible to do so, and we will only fail to recognise these facts in the face of the criticism by the public and its institutions.

There are two types of risks, namely those which may result in a loss or a gain - speculative risks - such as the potential collapse of a substandard bridge still in operation, and pure risks such as the potential damage caused by impact, fire, flood or earthquake.

When dealing with risks - to life, environment or property - four components should be considered:

a) Identification of risk exposure

In the management of bridges an inventory must be developed of all forms of possible loss. This will cover all stages in the life cycle of the bridges from the preliminary design to the final demolition.

The most important risks are those to which passing traffic is exposed - from defective or inferior guardrails to structural collapse; but such potential traffic hazards, as narrow bridges, limited headroom, inferior bridge deck surfacing and frequent, rapid icing of the bridge surface also call for serious consideration.

The recent OECD report "Evaluation of Load Carrying Capacity" deals with the questions of substandard bridges exposed to increasing traffic loads and also with the complications and risks involved when dealing with the growing number of exceptional heavy vehicles.

Environmental risks are limited in bridge management but should not be ignored. Taken in its widest sense accidental hindrance of traffic crossing a bridge may temporarily cause widespread environmental damage if heavy traffic must plough through areas not able to cope with it.

Inferior structural maintenance - routine as well as special maintenance - will cause potential loss of investments or initiate major repairs. Disregard or neglect of deterioration or faults in structural members may create risk of collapse.

b) Quantification of risk frequency, severity and amount

The measurement of the potential consequences of the identified exposure includes how frequently will losses occur, how severe will these be and what total risk potential may be present.

Concrete information on these items is practically non-existent. The development and establishment of data banks including data from bridge inspection reports may create the necessary data basis enabling us to make more accurate assessments of the risks and thus by proper maintenance avoid potential losses as well as costly overkills.

c) Analysis and evaluation

An evaluation of the identified exposure and their potential consequences must be made in the light of the general policy of the affected bridge or road authority. A certain amount of risk must be accepted. It is well known to all bridge authorities that heavy traffic loads are not removed from substandard bridges simply by putting up traffic signs prohibiting heavy axles on such bridges, and the risks involved by doing so must be evaluated against the costs of replacement and the traffic importance of the road in question. Similar considerations must be made regarding bridges inferiorly maintained or with limited headroom or width, etc.

d) Selection of techniques

The elimination of all risks is the desirable, unattainable goal in bridge management, but some of the risks can be effectively reduced when the problems are analysed realistically.

The human error syndrome is possibly the most important factor in these endeavours. A wide discussion on this topic is outside the scope of this report, however, it should be stressed - as it indeed was in the OECD report on bridge inspection - that education and training of personnel at all levels, attractive employment and proper organisation are measures towards eliminating the predisposing error conditions, and that the implementation of these measures without doubt will reduce the risks significantly.

A technical description of the risks and suggestions on how to reduce them has been given elsewhere in this report and in previous OECD reports.

The maintenance policy must strike an acceptable balance between the residual risks and the human and economic efforts needed to confine them.

There is a freedom of choice in coping with the speculative risks (inferior inspection and maintenance, substandard load carrying capacity, inferior surfacing, limited headroom and width) and a limited choice with the pure risks (impact, fire, flood or earthquake). The latter risks can be reduced in design codes, but not eliminated.

IV.2.3 Economic aspects

When considering bridge maintenance from an economic point of view the life cycle of bridges must be considered.

Bridges are built, maintained and repaired, demolished and rebuilt. The time it takes to complete a full cycle is the lifespan of a bridge, the length of which can vary greatly depending on environment, quality of design and construction and quality of maintenance and repairs.

In a given environment (traffic loads, climatic conditions, etc.) bridge building and maintenance techniques offer a choice of quality levels within which the task can be performed. In some countries the engineer can to a certain extent choose the quality of construction and in others quality is specified mainly by design standards, but regardless of who sets the standards of quality this choice does relate to the total costs of construction and maintenance. Maintenance work usually allows a choice of quality levels provided that the requirements for safety of the user are not neglected.

The cost of higher quality varies: where it is a matter of better materials or workmanship the cost is often small, but where it involves, for instance, an additional protective system then the cost may be much larger. If money were no problem, the normal strategy would be to build, maintain (and repair) bridges so they last almost indefinitely.

Evaluation of construction and maintenance policies

Within the cycle of building, maintaining and replacing bridges many decisions must be made. There is usually a choice of type of bridge, sometimes a choice of quality of construction and usually a choice of both qualities and techniques for maintenance. How are the alternatives to be evaluated?

To enable us to consider an answer to this question the significance of future costs in maintenance, repairs and replacement has to be evaluated, e.g. are bridges maintained because it is desirous to push the costs of replacement into a distant future?

Future costs are of lower present value or consequence than present costs, and this is certainly a fact for all road authorities.

To quantify this the concept of discounting is employed, i.e. how to express a future cost $c_{T=t}$ by its present equivalent, $C_{T=0}$:

$$C_{T=0} = e^{-r \times t} \cdot C_{T=t} \quad \text{(continuously}$$

or

$$C_{T=0} = (1 + r)^{-n} \cdot C_{T=t} \quad \text{discretely)}$$

where t is a given time span, n the equivalent number of time periods, and r is the time preference rate or discount rate. The term "interest rate" is here intentionally avoided as interest rate may only be a subconcept to the discount rate.

Figure IV.1 shows the effect on future costs of using the concept of discounting. The abscissa gives the numbers of years ahead of present time when a future cost shall be carried (defrayed), and the ordinate axis gives its present value in percentage of the future cost. The relationship is shown for various discounting rates 1 per cent, 2 per cent, 3 per cent and up to 15 per cent. Although not absolutely necessary, it is assumed – to avoid confusion – that we operate at the same cost level, i.e. no inflation! The discount rate does therefore not contain the rate of inflation.

It is important to realise, when looking at Figure IV.1, that the two limit values, the zero point (t=o) and the asymptote, $\frac{C_{T=0}}{C_{T=t}} = 0$, are undoubtedly true. How to reach from from the zero point to the asymptote can be discussed, but the choice made here is a widely accepted one.

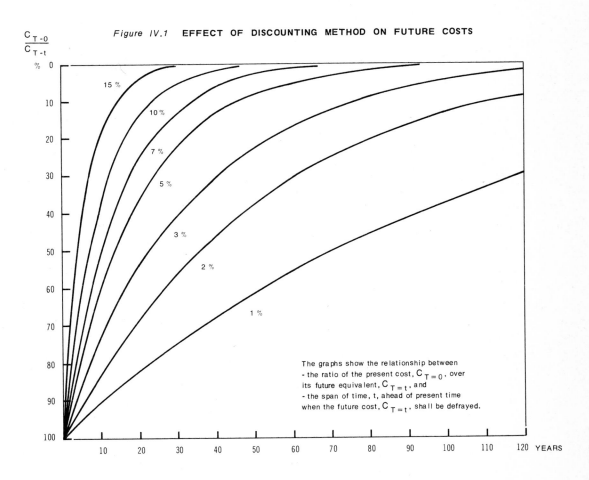

Figure IV.1 EFFECT OF DISCOUNTING METHOD ON FUTURE COSTS

The graphs show the relationship between
- the ratio of the present cost, $C_{T=0}$, over its future equivalent, $C_{T=t}$, and
- the span of time, t, ahead of present time when the future cost, $C_{T=t}$, shall be defrayed.

The concept of discounting can be an aid in decisions on actual problems such as:

- Are repair works that cost 40 per cent of a replacement economical if the repair postpone the replacement for 15-20 years?

The answer from Figure IV.1 is yes, if the present discount rate is above 3 per cent without inflation.

The consequences of the discount rate concept are perhaps best illustrated through an example, given in the Appendix D which illustrates the consequences of varying discount rates.

The example shows the extent by which the economic horizon is shortened by discounting, particularly where discount rates are high. A saving in the initial cost thus becomes of greater significance than a similar amount saved in the future cost of maintenance and replacement. While there is a clear benefit from selecting a type or design of a structure which will give least overall costs, the initial costs should not be reduced by lowering the quality of construction below accepted standards for good durability. A reduction in quality will certainly lead to increased maintenance and earlier replacement. Both these factors are difficult to quantify, but in work of lower quality there is a tendency for the rate of deterioration to be greater than expected. If the policy of a road authority is to accept low standards, this would result in a heavy burden of future expenditure.

Discount rates

The discount rate is the rate at which the future should be discounted as compared with the present. In some countries an approved discount rate must be used. For instance, in the United Kingdom, decisions on the choice of type of bridge or on maintenance techniques must be based on a discount rate which represents the opportunity cost of capital (currently 5 per cent) whereas the appraisal of new projects, which are not subject to market forces and where there is a greater risk of appraisal optimism, must be based on a higher rate (currently 7 per cent).

In other countries discount rates may be determined according to circumstances, such as in accordance with how much money is available to fulfil the total present requirements. If the requirements for maintenance exceed the total funds available for this purpose, the most needed maintenance work is given priority over maintenance work that can be deferred. This act translates into an increase of the discount rate. Should appropriation of funds be sufficient to meet all requirements then the concept of discount rate has only an academic interest. Chapters II and III of this report clearly state, however, that only very few, if any, OECD countries are so fortunate.

The availability of funds to meet maintenance requirements is a political matter - it may be a choice between social welfare and maintenance of public roads - but this report should assist in striking the right balance between appropriations and a reasonable level of bridge maintenance. It is generally not reasonable to build bridges that have a life of only about 20 years, whereas some countries have difficulty in building bridges at reasonable costs that will last much more than 50 years, if exposed to harsh climatic conditions and the use of de-icing salt. So from a technical point of view discount rates between 3 per cent and 7 per cent seem in general reasonable.

It should be stressed that this recommendation is based only on a technical assessment of bridges in the harsh environment of some OECD countries. Traffic considerations as well as a milder climate may lead to different conclusions.

It is important to underline that the discount rate in the construction phase and the discount rate during maintenance and repairs should be comparable (the same). During new construction – say of a new motorway – funds are often available to keep a high standard in the design and execution, which translates into a low discount rate. During the following maintenance period, there is often scarcity of funds, which imposes a higher discount rate upon maintenance works.

A division of discount rates is sometimes used in distinguishing between major bridges and those of normal size, as design, quality of construction and maintenance operations for major bridges in general are carried out to a greater degree of perfection than for ordinary bridges. This attitude becomes self-evident when the economic and traffic risks of inferior maintenance of major bridges are analysed objectively.

Use of discounting procedures

In using the discount concept it is most important that realistic estimates are made of the extent and cost of maintenance.

Allowance must be made for increases in real costs in the future such as those necessitated by piecemeal work at nights (through growth of traffic) and by the large labour content of maintenance costs. Account should be taken of all consequential costs such as those due to traffic diversion and increasing delays. There is also a strong tendency to underestimate the extent of future maintenance and this should be corrected on the basis of experience.

For replacement of the bridge, paragraph I.5.2 states that the cost of reconstruction of a bridge in service may be 1.5 to 8.0 times that of the initial cost of construction. If cost data for discounting is based upon estimates which take account of all the factors discussed above, then the comparison of present costs will provide a valid basis for making bridge management decisions. These could include the choice of type of bridge, quality of construction (within the limits allowed by design standards and by the generally accepted objective of a design life of not less than 100 years) and policies for maintenance and future replacement.

If an approved (fixed) discount rate is used for all bridges, full allowance should be made for any differences in the cost of maintenance or replacement including the cost of delays or diversions of traffic both over and under the bridge. The costs will thus reflect the amount of traffic carried and the importance of the bridge.

This is obviously the ideal way of using the discount concept. Estimates of such future costs – even the order of such costs – are, however, difficult to obtain, and the differentiation of discount rates, as is described above, represents therefore an alternative tool in bridge management, however crude and inaccurate it may seem to be.

IV.2.4 Traffic aspects

Road bridges pose peculiar traffic problems because they are the more rigid and relatively permanent elements of the route. Whereas widening and strengthening of the roads themselves can be effected concurrently with the increase in traffic volume and load (usually with traffic restrictions but rarely with serious hindrance to traffic), the same benefits can seldom be achieved at bridge crossings without total replacements. Such replacements are both costly and technically complicated if the road must remain open during the replacements. Many roads possess therefore substandard bridges regarding traffic loads as well as traffic capacity and traffic risks.

These facts must enter into the maintenance policy. When a bridge constitutes a traffic hazard we must confine this hazard by more vigilant inspection and more vigorous maintenance. It is therefore necessary to back up such policy by allocating funds to such bridges possibly to the detriment of the maintenance of bridges without such risks.

The discount concept can be employed in this policy as well as in the evaluation of - or the distinction between - maintenance operations on bridges serving important traffic arteries and those serving more humble roads.

A simple division of bridges into three categories in accordance with their traffic importance, traffic volume and traffic risk could serve the purpose of funnelling resources into bridge maintenance where it will serve the traffic best by having a low discount rate for bridge maintenance on important traffic arteries and a higher discount rate on less important roads.

A variation of the discount rate of only 1-2 per cent in such a differentiation can over a long period influence the life span of the bridges. Figure IV.1 indicates that such a variation in the discount rate can ultimately lead to a difference in life spans of about 20 years.

Whether such a policy is an economical one is ultimately a political issue, but as long as bridge maintenance must compete with other necessary demands upon the taxpayers' contributions, the differentiation of the discount rates is one way of letting the more important roads and bridges gain an economic advantage at the expense of the less important ones.

IV.2.5 Technical aspects

All materials used in bridge building deteriorate. Some forms of deterioration (e.g. rusting of steel) can be effectively contained, if appropriate preventive measures are applied, while other forms of deterioration (e.g. weathering of concrete, asphalt pavement and waterproofing) are less amenable to preventive maintenance and must be dealt with primarily in the design and construction phase.

The problems of concrete durability often receive insufficient attention. If concrete is kept dry the material may last forever, and if the concrete is dense without cracks, with the correct amount of entrained air, etc. it will be able to sustain for a long time a wet and harsh climate with numerous freeze-thaw cycles including the action of de-icing salt.

However, although durable concrete can be recognised in the laboratory, it is not always produced when constructing bridges.

Part of the problem is the attitude to the material "concrete". It is generally regarded as a very robust material which is handled roughly (and cheaply) during production, when the fact of the matter is that the durability properties are hypersensitive to the accuracy and solicitude with which the concrete (and its aggregates) are mixed, transported, placed, compacted, cured and treated during hardening until it has reached final maturity.

The demand for durable concrete, with the qualities needed in bridge building, is only a fraction of the total concrete production. There exists therefore no economical incentive amongst concrete producers to change the above-mentioned attitude. Bridge authorities must specify the strictest measures of control to counteract this general attitude - and pay for it - if exposed concrete structures are to achieve a higher degree of durability.

The development of more durable concrete is under way, and should be encouraged. A wider use of flyash and other waste materials resulting from the increasing environmental demands upon industrial production may hold promises of using such waste while producing more durable concrete.

The present lack of durability in concrete has led to a whole series of remedial measures to be applied to exposed concretes surfaces with the aim of arresting further deterioration. Care should be taken when selecting such a remedy both with regard to its properties as well as to its expected effectiveness. Regarding the former, it is essential that the concrete surface is kept open for interior water to escape by evaporation.

The above example of concrete durability illustrates the many and complex technical aspects in bridge maintenance. There are - apart from concrete - a number of structural elements and materials which are lacking in durability, for example:

- Pavement and waterproofing. The constant development in this field has given more durable but also more complex and costly systems of pavement/waterproofing with a high sensitivity to any variation in the quality of workmanship and climatic conditions during the construction.
- Bituminous joints. Such durable joints are still wanting. At present only the frequent replacement of ordinary bituminous joints can meet the maintenance needs.
- Expansion (deck) joints. The increasing impacts from traffic have made design of durable expansion joints difficult. The present development is towards joints that are very robust but also very expensive.

IV.3 MAINTENANCE ENGINEERING

Through this Chapter several important means to implement a bridge maintenance policy have been mentioned. In the following the significance of proper bridge design, of easily accessible information and of training and research is touched upon, however, there are many other important factors being of general as well as of more special nature. While existing bridges may include steel, cast iron, concrete, masonry, brick and timber structures, all of which have different maintenance problems, it is the maintenance of modern steel and concrete structures which poses the more significant engineering challenges in the overall bridge management field.

IV.3.1 The significance of bridge design

Bridge design must consider the total costs of the structure throughout its life span, including maintenance, repairs and replacement, and it should be emphasized that feeding maintenance experience back to the bridge designers is imperative for the reduction and improvement of future maintenance operations. Most maintenance considerations in bridge design can be applied without increasing the initial cost of the structure and some - perhaps very obvious - features must be adopted in the design to serve this purpose:

Accessibility

Accessibility to all parts requiring inspection and maintenance is a must. These parts are in fact all structural parts above ground plus some under water or underground such as foundations, drainage, dewatering installations, geotechnical check points, etc.

Stairs, ladders or passable slopes on inclined planes are obvious features. Sometimes just a path will do. The use of more advanced inspection and maintenance equipment such as travellers and hydraulic lifts should be made possible and the flow of traffic must be considered during operations of inspection and maintenance. Trespassers should, however, be discouraged from making use of such improved accessibility (by steel doors, railings, etc.). Especially sensitive points such as bearings and expansion joints must be easily accessible for visual inspection.

Maintenance considerations in bridge design

Bridges should be designed so that the need for maintenance is minimised or - if possible - even avoided, with due regard to the possible increase in construction costs, cf. Chapter IV.2.3: Economic Aspects.

The purpose of preventive maintenance is to avoid, contain or restrict progressive deterioration. Preventive maintenance starts with the design.

The choice of materials is important: weathering steel, stainless steel, light alloys and plastics are rarely used, although these materials apparently possess long durability characteristics. Radical changes of materials for structures of long life spans such as bridges should be investigated thoroughly before application is decided, and a conservative approach should be adopted so that changes only come with substantially proven techniques.

Many of the features in experienced bridge design are directly aimed at the goal of high durability. However, some very obvious and inexpensive arrangements are often overlooked.

The main agent in the deterioration process is water, so maximum attention should be given to ridding the bridge decks of water as speedily as possible.

Maintenance work and costs can be reduced significantly when the design of bridges includes considerations such as:

- water, snow and rubbish must be quickly and easily removed from bridge decks;
- water seepage through pavement and waterproofing must be zealously checked;
- the use of vulnerable construction elements such as expansion and construction joints and bearings should be restricted even if the initial costs are increased. This consideration may lead to the conclusion that certain bridge designs such as those involving extensive use of prefabricated elements may incur higher maintenance costs although the initial construction cost may be lower than that of other designs;
- the installation of water, gas, heating and other supply pipes, lamp posts and drainage or sewage pipes should be avoided or restricted, however, should such installations be accepted or expected during the life time of the bridge, the bridge design should consider the risks involved as well as the added maintenance operations and the costs hereof.

Replacement of deteriorated elements

As certain bridge parts deteriorate faster than others, an application of partial replacement options could prove advantageous. If the more exposed parts are made replaceable, it is important that the bridge design considers all steps in the future replacement procedure.

The replacement options could include:

- bearings
- expansion joints
- parapets
- railings
- installations (culverts, drainage pipes, etc.)
- pavement, waterproofing
- bituminous joints

In this connection, careful consideration should be given to the fact that many bridges - especially large bridges - of modern design use the bridge deck as an integral part of the overall static function, e.g. bridges built by segmental methods. Every effort must be made to prevent any deterioration of such bridge decks, because their replacement cannot be contemplated. In fact a number of modern bridge design concepts integrate by means of advanced computer techniques the overall structure in a complicated static pattern that reduces the possibility of future replacement of structural members without serious complications and high costs.

IV.3.2 **Technical assessment for maintenance**

When inspection reveals deterioration or defects in a structure, it is first necessary to determine whether remedial treatment is necessary and then to decide what measures should be taken and when.

Several matters need to be considered: does the defect pose any risk to the safety of users or to the structure? Will it reduce load-carrying capacity? Will the condition of the structure worsen? Are there any treatments available which will effectively remedy the situation? Can the cause of the trouble be removed? In considering these matters the need for early preventive action to avoid more difficult repair work later must always be emphasized. There are, however, also repairs which can be and should be postponed, e.g. painting of steel structures or repairs of deteriorated concrete should not be contemplated before certain stages in the deterioration process have been reached.

The criteria at present used in different countries have been described in II.4.2 and the catalogue of defects used in France is referred to in Annex A. At present the answers to the questions listed above depend very much upon the individual experience of the engineer. While it is likely that such experience will continue to provide the basis for many decisions on the multitude of tasks that comprise maintenance, there is a real need for the collective experience on bridge maintenance to be assembled and evaluated on both a national and international basis. The establishment of maintenance manuals could be very useful in this respect. The increasing number of sophisticated structures built a decade or more ago and now beginning to need maintenance lends weight to the need for wider studies.

IV.3.3 **Maintenance information systems**

A maintenance organisation should systematically examine and record the effectiveness and condition of maintenance work done in previous years. This will show the shortcomings in relation to both materials and workmanship and thus indicate where improvements are needed. The records will enable the long-term durability to be assessed and so allow a better evaluation to be made of different materials and methods.

Various electronic data processing (EDP) systems or data banks containing bridge data have been established in some countries and further development of such data banks including data from bridge inspections is at present under way. The ideal bridge data bank should contain the following information for all bridges:

- Identification: Unique structure number related to road network.
  Map reference. File number.
- Administrative data: Owner
  Maintaining authority
  Agreements, Way leaves, etc.
  Statutory mains, services, etc.
  Year constructed
  Design certificates and records
- Crossing data: Over or under road, railway, water, etc.
  Load carrying capacity
  Clear width
  Headroom
  Cross reference to crossing road

- Structural data: Structure: road bridge, retaining wall, culvert, etc.
　　　　　　　　　　Form: number and length of spans, skew, etc.
　　　　　　　　　　simply supported or continuous, frame, arch
　　　　　　　　　　Construction: solid slab, beam and slab, etc.
　　　　　　　　　　　　　　reinforced concrete
　　　　　　　　　　　　　　prestressed concrete
　　　　　　　　　　　　　　structural steel
　　　　　　　　　　　　　　masonry or brickwork
　　　　　　　　　　　　　　foundations
- Ancillaries:　　　Expansion joints
　　　　　　　　　　Bearings
　　　　　　　　　　Parapets
　　　　　　　　　　Waterproofing
- Inspection data:　Date last inspected
　　　　　　　　　　Summary of condition
　　　　　　　　　　Evaluation of executed maintenance and repairs
- Particular problems: Maintenance costs
　　　　　　　　　　Durability of components
　　　　　　　　　　Paint systems
　　　　　　　　　　Feedback on specific detail

The scope of a data bank must be tailored to the precise needs of the authority because it will depend on the reliability of the data fed into it and the gathering and storing of information is a costly business. Proper consideration should be given to the selection of the computer hardware and software having regard to who will actually operate it and how often. Microprocessors, optical mark readers, etc. facilitate direct input to the computer. Whilst computerised data processing is a powerful tool for the engineer, he must strike a balance between collecting a vast amount of data that will never be used and not having enough to justify the use of the techniques. Realistic provision must be made both in manpower and money for the initial input to set up the system and afterwards for keeping it up-to-date and running it.

An example of a bridge data bank system that contains such features is given in Appendix C.

IV.3.4 Training and research

Training is needed at two levels. The professional (or registered) engineer will usually find it difficult to keep his knowledge up-to-date on the testing techniques, materials and maintenance procedures which are available to him. Short up-dating courses addressed by specialists in these fields can be most valuable and should be held at intervals of about 5 years.

The workmen in the maintenance team will often be using materials and techniques of which they have little experience. The need for precise batching of materials, for cleanliness, for observing limitations on temperature and moisture tolerances and many other factors are not always appreciated. Practical courses are desirable for all maintenance teams with additional short courses whenever new materials or techniques are introduced. An important aspect of these courses should be health and safety. All maintenance teams should understand the precautions needed in maintenance work. This should include the use of protective clothing, helmets, breathing masks and eyeshields; working on elevated structures, on scaffolding, in confined places and over water; traffic hazards and the use of proper control and warning signs; responsibilities to fellow workmen and road users generally and national regulations on health and safety.

With regard to research there are four main requirements. Firstly the need to investigate the effectiveness of present methods of maintenance. These investigations should be made in the context of the life expectancy of the structure being maintained and the effectiveness of the work in preserving the load-carrying capacity of the structure and ensuring the safety of users.

The second need is to develop criteria which will enable the maintenance requirements to be evaluated on a more consistent and systematic basis.

The third need is to develop improved materials and techniques for bridge maintenance, with particular regard to the problems now arising in some modern structures.

The fourth need is for a study of the economics of bridge maintenance based upon actual costs for bridges of various ages and types.

REFERENCE

BERISSI R. and CANTEGREIL J.P. Le fichier des ouvrages d'art. Bulletin de liaison des Laboratoires des Ponts et Chaussées, No. 83, Paris, May-June 1976.

Chapter V

MAINTENANCE RESEARCH

V.1 RESEARCH CONSIDERATIONS

It has been pointed out in previous Chapters of this report that the intelligent management of a total highway system involves the processes of design, construction, inspection, assessment of condition, maintenance, and replacement when necessary. The above pertains to pavements as well as the structures on a highway. It has also been noted that all the above processes are interconnected and relate to each other, and that there must be open dialogue between the persons responsible for the various functions. It is therefore somewhat difficult to clearly separate out research considerations that pertain only to bridge maintenance, which is the subject of this report.

After reviewing current and planned research in paragraphs V.2 And V.3, this Chapter presents recommendations on research needs. The importance of bridge management needs indicated at the end paragraph V.4 should be stressed, i.e. accurate maintenance cost accounting, quantified bridge condition assessment and lastly general consideration.

V.2 CURRENT RESEARCH

The listing of research studies presently under way along with the name of the organisation, the amount of funding in thousands of dollars, and principal investigator are presented in paragraph V.5. A great deal of structural research has of course been performed and reported in recent times, and a list containing a considerable number of completed research study reports having to do with some aspect of bridge maintenance is presented in section V.6.

The studies can be grouped in a general manner, dealing with:

a) Concrete bridge decks and attendant waterproofing and reinforcement corrosion protection;
b) Damaged concrete repair caused by accidents or natural causes;
c) Prestressed concrete member repair;
d) Old bridge strengthening;
e) Bearing and joint repair and maintenance;
f) Various aspects of metal cleaning and painting;
g) There is also some work under way on the general subject of bridge maintenance management.

This list is by no means comprehensive, and is in fact a very limited account of the bridge research in progress that might have an influence on bridge maintenance.

V.3 PLANNED RESEARCH

Little information made available to the Group dealt with future bridge maintenance research; the following is taken principally from the reply of France. Some of the studies are very similar to research already in progress, in that they would deal with the characterisation of the corrosive environment surrounding bridges. Other research would pursue the methods of concrete crack repair, and repair of damaged bridges in general. The above would also include surface treatments of various kinds to protect existing concrete from further moisture related damage.

One of the studies would have as its object the maintenance data treatment. That is, it would be a paper study dealing with the financial and economic part of bridge maintenance, sorting this information in such a manner that it would be suitable for analyses that point to the relationship between the damage and its causes.

Masonry repair and maintenance is the subject of another study. This is both a technical study as well as an economic one to define the best methodology of masonry maintenance. In a similar manner, timber bridges will be treated in yet another study.

Some research will be continued in new studies on bearings such as reinforced elastomers, dealing specifically with their long-term durability. There will also be continuation of research on the best wearing course for bridge decks.

A very comprehensive research study is being planned in West Germany, combining industry, the academic communities and the bridge owners. The study will deal mostly with the long-term behaviour of prestressed concrete bridges. It will deal with the fundamentals of concrete behaviour when subjected to the influences of freezing and thawing, as well as atmospheric damage, and will seek to shed light on the fundamental parameters involved in long-term damage. The study is to also involve physical testing of laboratory specimens, field observations and theoretical work, all aimed at forecasting behaviour. It is to begin in the spring of 1980 and continue for several years.

V.4 RECOMMENDATIONS FOR FUTURE RESEARCH

Chapters II and III of this report outline the present situation regarding maintenance policies and techniques existing in the member countries, and Chapter IV outlines a desirable proposed bridge maintenance policy. It is apparent that in order to close the gap between the present practice and that which might be an optimum desirable practice, a great deal of knowledge about all aspects of bridge maintenance must yet be gained and disseminated. It is therefore urged that carefully planned and executed research studies be continued and undertaken so that the desirable goal of a long, trouble-free bridge life can be approached.

The suggested research can be broadly separated into physical research dealing with the behaviour of structures and their elements in specific environments subjected to loads, and the non-physical investigations dealing with costs, practices, and policies of bridge maintenance.

A partial list of physical maintenance research needs is as follows:

1. Repair of concrete - this work would involve all types of crack repair, including the development of suitable materials that could withstand the motions of an "active" crack, as well as methods and techniques for the rejuvenation of generally deteriorated concrete.

2. Waterproofing and corrosion protection of reinforcement – this would include continuation of the search for an economic substitute for de-icing salts, as well as external and internal methods of concrete protection.
3. Improvement of moveable joints – the maintenance of joints is a never-ending problem, and their elimination where structurally acceptable in new designs as well as in existing bridges should be explored. The development of newer, more trouble-free joints should be continued. Of particular interest is the behaviour of "buried" expansion joints.
4. Post-tensioned bridge maintenance and strengthening – this type of structure often incorporates many joints with the attendant waterproofing problems, along with cables, ducts, and anchorages that are subject to deterioration, raising the possibility of very serious structural problems. Research should therefore also focus on better injection and grouting materials and technology.
5. Live load monitoring – since some maintenance requirements are directly related to the live loads using the bridges, a better, more comprehensive programme of load monitoring is desirable.
6. Fatigue crack repair – efforts should continue on methods of repairing or arresting fatigue cracks in steel bridges, but research should also include the development of preventive measures and techniques for the lengthening of the fatigue life of bridge members.
7. Corrosion protection of steel structures – efforts should continue to develop long-lasting, environmentally safe and economical paint systems. Other corrosion inhibiting materials and manufacturing processes need to be encouraged and evaluated. Of particular interest at present are the problems being experienced in North America with weathering steels, which under some environmental conditions seem to continue to corrode. The waterproofing of cables and cable connections also continues to be of great concern.

Those research needs that relate to cost accounting, maintenance practices and strategies, and which would lead to an overall bridge management approach, are detailed as follows:

1. Accurate maintenance cost accounting – there exists a definite need for the setting up of studies that separate out the specific costs associated with various maintenance tasks, such as painting, cleaning of gutters, drains, etc., crack repair, tightening of bolts, joint and bearing repair, and any other relatively routine maintenance, so that an intelligent assessment can be made as to how best to use the resources to assure the optimum life and safety of our structures.
2. Long-term behaviour studies – it is very desirable to have accurate knowledge about the material behaviour of structures over a long period of time, as well as the overall long-term structural behaviour of the entire bridge. Reference here is made to the study of corrosion rates in various environments, changes in concrete properties, crack pattern development, fatigue crack growth, relaxation in prestressing force, settlement rates, scouring, timber deterioration, and any number of other long-term phenomena. Of interest especially to the subject of this report would be any methodical evaluation of the long-term effectiveness of various repair techniques, such as for instance the durability of certain epoxies used for concrete crack repair.
A note of caution is in order here: Whereas it may on the surface seem to be desirable to set up very comprehensive data gathering schemes and lists of

various kinds, it will soon be apparent that a deluge of undigested data can quickly result which negates any good that might have resulted from the study. A great deal of prudence is to be exercised when setting up any data gathering scheme.

3. Quantified bridge condition assessment - the need continues to exist for better means of judging the load carrying capacity and general serviceability of a bridge at any one time, as well as better ways of accurately forecasting the remaining life of a structure. Development of tools and measuring devices, stress counters, even better analytical methods, should be encouraged in order to remove some of the empiricism and "seat-of-the-pants" engineering connected with this subject.

4. Maintenance effectiveness and strategies - studies are needed in which certain controlled variations of maintenance strategies and practices are made, such as the time variations of some repairs, or even the complete neglect of any maintenance, so that the relative economics can be optimised with serviceability and safety. Such studies would help to determine annual maintenance costs and replacement rates of bridges.

5. General considerations - in summary, although a number of specific areas of needed research have been mentioned, the greatest need seems to be the development of a true cost accounting of the separate maintenance tasks performed, so that maintenance costs can be related back to design considerations and to the specific forces creating the damage.

A further important need is for the development of the methodology of bridge maintenance so as to provide improved criteria for assessing maintenance needs and to provide guidance on determining the optimum maintenance strategy. Due regard must be given to safety, economic, traffic and technical aspects of maintenance.

V.5 LIST OF SOME ONGOING RESEARCH

Feasibility Evaluation of the Implementation of Concrete Polymers for
   Bridge Deck Applications
   Oklahoma University - Locke $ 80K

Evaluation of Bridge Deck Repair and Protective Systems
   Colorado Department of Highways - L.B. Steere

Dynamic Response of Bridges to Seismically Induced Ground Vibration -
   Phase 2 - Monitoring and Maintenance
   California Department of Highways - R.E. Davis   49K

Bridge Deck Restoration Methods and Procedures
   California Department of Highways - J.B. Poppe   179K

Heating Coils in Pavements and Bridge Decks
   Kentucky Department of Transportation - J.H. Havens   42K

Implementation of Polymer Concrete Technology for Highway Repair and
   Maintenance in the State of Georgia
   Georgia Institute of Technology - D.J. O'Neil   65K

Corrosion Study of Bridge Decks
   Nebraska Department of Roads - J. Anderson   25K

Concrete Bridge Deck Repairs Using Injected Epoxy Resin
   Iowa Department of Transportation - E.J. O'Connor

Thin Polymer Concrete Overlays
   New York State Department of Transportation - Mediatore   44K

Bridge Deck Patching New Jersey Department of Transportation - Margerum	69K
Evaluation of Bridge Patching Materials Massachusetts Department of Public Works - Romano	140K

Evaluation of Demage and Methods of Repair for Prestressed Concrete
   Bridge Members
   George O. Shanafelt, Olympia, Washington
	59K

Field Evaluation of Microwave Patching System Illinois Department of Transportation - P.G. Dierstein	59K
Cathodic Protection for Reinforced Concrete Bridge Decks Texas State Department of Highways - T. Kenner	23K

Bridges on Secondary Highways and Local Roads, Rehabilitation and
   Replacement
   University of Virginia - Kinnier
	120K

Cathodic Protection of Concrete Bridge Structures
   Corrosion Engineering and Research Company,
   Concord, California - W.J. Ellis
	225K

Maintenance Methodology of Bridges L.C.P.C., Paris, France - M. Bastet	
Criteria of Site Aggressiveness L.C.P.C., Paris, France - M. Raharinaivo	

V.6 BIBLIOGRAPHY

"Physical Facility Maintenance - Structures", International Bridge Tunnel and Turnpike Association, J.P. Zitelli, November 1971.

"Concrete Overlays for Bridge Deck Repair", Highway Research Board, Washington, D.C., Texas Transportation Institute, H.L. Furr, L.L. Ingram, 1972.

"Repair of Hollow or Softened Areas in Bridge Decks by Rebonding with Injected Epoxy Resin or Other Polymers", Kansas DOT, Topeka, Kansas, F.W. Stratton, B.F. McCollom, 1972.

"Administration of Bridge Inspection", Action Guide Series, Volume XI, National Association of County Engineers, Cedar Rapids, Iowa, J.L. Campbell, July 1972.

"Bridge Deck Deterioration: A Summary of Reports", Texas Transportation Institute, Furr, Ingram II, Moore, Swift, December 1972.

"Polymer Resins as Admixtures in Portland Cement Mortar and Concrete", FHWA, Washington, D.C., R.G. Pike, R.E. Hay, October 1972.

"Bridge Railing Repair Coating with a Paraffin Base Sealant", FHWA, Washington, D.C., J.C. Patten, April 1972.

"California Answer to the Bridge Paint Problem", Public Works Journal Corp., Ridgewood, New Jersey, December 1973.

"Linseed Oil Retreatments to Control Surface Deterioration of Concrete Bridge Decks", Illinois DOT, D.D. Fowler, R.H. Mitchell, March 1973.

"Underwater Bridge Inspection", Civil Engineering, Volume 43, No.3, ASCE, J.J. Powers, T.J. Collins, March 1973.

"An Instrument for Detecting Delamination in Concrete Bridge Decks", Highway Research Record N 451, Transportation Research Board, Washington D.C., Moore, Swift, Milberger, 1973.

"Detection of Bridge Deck Deterioration", Highway Research Record N 451, Transportation Research Board, Washington, D.C., W.N. Moore, 1973.

"Special Report on Mobile Scaffold for Painting Structural Steel Bridges", FHWA, Washington, D.C., M.H. Anderson, December 1973.

"Halting Deck Deterioration on Existing Bridges", Civil Engineering, Volume 43, No.10, ASCE, New York, NY., G. Dallaire, November 1973.

"A Review of Field Applications of Fibrous Concrete", Transportation Research Board No.148, Washington D.C., W.A. Yrjanson, 1974.

"Bridge Replacements with Precast Concrete Panels", Transportation Research Board No.148, Washington, D.C., Biswas, Iffland, Schofield, Gregory, 1974.

"Steel-Fiber-Reinforced Concrete", Transportation Research Board No.148, Washington, D.C., W.L. Gramling, T.H. Nichols, 1974.

"Techniques for Evaluating Reinforced Concrete Bridge Decks", Ohio State University, Columbus, Ohio, Ohio Transportation Engineering Conference Proceedings, Volume 28, J.R. Van Daveer, April 1974.

"Effectiveness of Attached Bridge Widenings", California DOT, Sacramento, California, R.V. Shaw, C.F. Stewart, March 1974.

"Debris Removal from Concrete Bridge Deck Joints", Texas Transportation Institute, Texas A&M University, College Station, Texas, No. 12-F, W.M. Moore, September 1974.

"Experimental Cathodic Protection of a Bridge Deck", Transportation Lab., California DOT, Sacramento, California, R.F. Stratfull, FHWA-RD-74-31, January 1974.

"A Survey of Distress and Debris in the Joints of Pan-Formed Bridges", Texas Transportation Institute, Texas A&M University, College Station, Texas, Moore, Swift, Furr, January 1974.

"Manual for Maintenance Inspection of Bridges", AASHTO, Washington, D.C., 1974.

"Lighter, Orthotropic Design Replaces Deteriorated Deck", Engineering News-Record, McGraw-Hill, Inc., New York, N.Y., Volume 192, No.19, May 1974.

"Causes of Bridge Pier Staining", Arkansas University, Fayetteville, Arkansas, S.I. Thornton, C. Springer, February 1974.

"Problems of Repairing and Maintaining Reinforced Concrete Bridges", Avtomob Dorogi, Moscow, U.S.S.R., G.N. Borodin, I.A. Khazan, 1974-02.

"Polymer Concrete Overlay Test Program", Oregon DOT, Salem, Oregon, Jenkins, Beecroft, Quinn, FHWA-RD-75-501, November 1974.

"On the Possibilities of Applying Epoxy Resins in Hydraulic Concrete Repair Problems", A.M. Pailere, Y. Rizoulieres, Y. Lazzeri, Bull. Liaison Labo. P. et Ch., Paris, November-December, p. 111, 1974.

"Epoxy Paints - Their Use for Coating Steel and Concrete Structures", A.M. Serres, Bull. Liaison Lab. P. et Ch., Paris, November-December, p. 129, 1974.

"Cracking of Voided Post-tensioned Concrete Bridge Decks", Ministry of Transportation and Communications, Research and Development Division, Ontario, Canada, P. Csagoly, M. Holowka, 1975.

"Evaluation of Steel Bridge Inspection Instruments: Acoustic Crack Detector (ACD)/Magnetic Crack Definer (MCD)", Federal Highway Administration, Washington D.C., A.M. Lizzio, D.S. Nelson, FHWA-RD-76-502, January 1976.

"Open-Grid Decking Material 'Recycles', Old Bridges", Rural and Urban Roads, Scranton Publishing Company, Chicago, Illinois, Volume 14, N 3, March 1976.

"Research on Maintenance on Concrete Structures", Public Works Journal, Ridgewood, New Jersey, Volume 107, N 9, September 1976.

"Repair and Maintenance of Pavement Surfacings of Movable Metal Viaducts", P. Mehue, Bull. Liaison Labo. P. et Ch., Paris, p. 17, March-April 1976.

"Microwave Heating for Road Maintenance", Syracuse University Research Corporation, Syracuse, New York, No. SURC-TR-76-052, March 1976.

"Glulam Timber Could Figure Prominently in Booming Bridge Replacement Market", Construction Publishing Company, Arlington, Virginia, Volume 43, n. 41, February 1976.

"Measurement and Prediction of Preferential Icing Potential of a Bridge Deck", Transportation Research Record, Transportation Research Board, Washington, D.C., No. 598, 1976.

"An Evaluation of Concrete Bridge Deck Surfacing in Iowa", Office of Materials, Iowa DOT, J.V. Bergen, B.C. Brown, 1976.

"Corrosion Control of Steel in Concrete", Transportation Laboratory, California DOT, Sacramento, California, D.L. Spellman, R.F. Stratfull, November 1976.

"Resurfacing Bridge Decks", Civil Engineering Research Center, South Dakota School of Mines and Technology, Rapid City, South Dakota, W.V. Coyle, L.S. Iyer, FHWA-SD-77-608(72), November 1976.

"Modular Panels Speed Bridge Deck Repairs", Construction Methods and Equipment, McGraw-Hill, Inc., New York, N.Y., B. Lamb, Volume 59, No. 1, January 1977.

"The Evolution of the Anti-corrosion Protection of Metal Bridges", P. Mehue, Bull. Liaison Labo. P. et Ch., Paris, pp. 123-129, January-February 1977.

"System of Anticorrosion Protection for Cables of Suspension and Stayed Girder Bridges", D. André, Bull. Liaison Labo. P. et Ch., Paris, P. 134, January-February 1977.

"The Use of Paints with a High Zinc Content for the Anticorrosion Protection of Civil Engineering Structures", A.M. Serres, Bull. Liaison Labo. P. et Ch., Paris, p. 142, January-February 1977.

"Behavior of Epoxy Repaired Full-Scale Timber Trusses", ASCE Journal of the Structural Division, American Society of Civil Engineers, New York, N.Y., Avent, Emkin, Sanders, Volume 106, NST6, June 1978.

"Bridge Deck Restoration Methods and Procedures, Part III - Exposed Membrane Seals on Bridge Decks", California Department of Transportation, Sacramento, California, P.J. Jurach, I.R. Aarset, FHWA-CA-SD-77-15, July 1978.

"Cracked Structural Concrete Repair Through Epoxy Injection and Rebar Insertion", Kansas DOT, Topeka, Kansas, Stratton, Alexander, Nolting, FHWA-KS-RD-78-3, November 1978.

"Bridge Maintenance Products and Methods Used by the States for Deck Repairs, Structural Painting, and Various Other Activities", Maintenance Aid Digest, American Association of State Highway and Transportation Officials, Washington, D.C., MAD-16, February 1978.

"Influence of Bridge Deck Repairs on Corrosion of Reinforcing Steel", Battelle Columbus Laboratories, Columbus, Ohio, W.K. Boyd, NCHRP Research Results, Digest 85, 1978.

"Extending the Service Life of Existing Bridges by Increasing Their Load Capacity", Federal Highway Administration, Washington, D.C., R.H. Berger, FHWA-RD-78-133, June 1978.

"Bridge Deck Rehabilitation - Methods and Materials", Oklahoma DOT, Research and Development, Oklahoma City, Oklahoma, P.M. Ward, March 1979.

"Bridge Deck Delamination Study", Iowa DOT, Office of Materials, V. Marks, March 1979.

"Evaluation of Repair Techniques for Damaged Steel Bridge Members", Battelle Columbus Laboratories, Columbus, Ohio, H.W. Mishler, NCHRP, May 1979.

"A Study of High-Pressure Water Jets for Highway Surface Maintenance", IIT Research Institute, Chicago, Illinois, T. Labus, National Science Foundation, Division of Integrated Science and Public Technology, Washington, D.C., ISP76-12230, June 1979.

"Snow Removal and Ice Control Research", Transportation Research Board Special Report, TRB, Washington, D.C., No. 185, 1979.

"A Further Evaluation of Concrete Bridge Deck Surfacing in Iowa", Iowa DOT, Highway Division, Office of Materials, Ames, Iowa, March 1979.

"Resurfacing Concrete Bridge Decks", South Dakota School of Mines and Technology, Rapid City, South Dakota, W.V. Coyle, 1975.

"The Use of Polymer Concrete for Bridge Deck Repairs on the Major Deegan Expressway", Brookhaven National Laboratory, Upton, New York, Kukacka, Mediatore, Fontana, Steinberg, Levine, FHWA-RD-75-313, January 1975.

"Successful Electroshock Therapy for Deteriorated Bridges", Transportation Research News, Transportation Research Board, Washington, D.C., D.E. Robinson, No. 58, March 1975.

"Polymer Concrete for Repairing Deteriorated Bridge Decks", Transportation Research Record, Transportation Research Board, Washington, D.C., Kukacka, Fontana, Steinberg, No. 542, 1975.

"Bridge Pier Staining", Transportation Research Record, Transportation Research Board, Washington, D.C., S.I. Thornton, No. 547, 1975.

"Hot-Mixed Membrane for Bridge Deck Protection", Transportation Research Record, Transportation Research Board, Washington, D.C., Mellott, Saner, Prince, Kietzman, No. 554, 1975.

"Techniques for Reducing Roadway Occupancy During Routine Maintenance Activities", NCHRP Report, Transportation Research Board, Washington, D.C., No. 161, 1975.

"Considerations for Repairing Salt Damaged Bridge Decks", Journal of American Concrete Institute, Detroit, Michigan, C.F. Stewart, Volume 72, No. 12, December 1975.

"Assembly-Line Deck Speeds Bridge Renovation", Better Roads, Chicago, Illinois, Volume 45, No. 10, November 1975.

"Concrete-Polymer Materials for Highway Applications", Brookhaven National Laboratory, Upton, New York, Kukacka, Fontana, Steinberg, FHWA-RD-75-86, June 1975.

"Infrared Heating to Prevent Preferential Icing of Concrete Box Girder Bridges", Colorado Department of Highways, Denver, Colorado, R.G. Griffin, Jr., June 1975.

"Improvement of Mortars and Concrete Thanks to Epoxy Resins", A.M. Paillere, Bull. Liaison Labo. P. et Ch., Paris, p. 77, January-February 1975.

"Bridge Deck Repairs", NCHRP Research Results Digest, Transportation Research Board, Washington, D.C. No. 85, March 1976.

"A Status Report on Fiber Reinforced Concretes", Concrete Construction Publications, Inc., Addison, Illinois, Volume 21, No. 1, January 1976.

"The Preparation of Metal Surfaces Before Use: Substitute Abrasives", R. Lafuente, M. Persy, C. Rappenne, Bull. Liaison Labo. P. et Ch., Paris, p. 108, March-April 1977.

"Bridge Bearings", NCHRP Synthesis of Highway Practice, Transportation Research Board, Washington, D.C. No. 41, 1977.

"Handbook for Choosing and Applying Repairing Products for Concrete Structures", S.E.T.R.A. - L.C.P.C., 1977.

"Bridge Inspector's Manual for Movable Bridges", Federal Highway Administration, Washington, D.C., IP 77-10, 1977.

"Detection of Steel Corrosion in Bridge Decks and Reinforced Concrete Pavement", Iowa DOT, Ames, Iowa, V.J. Marks, HR-156, May 1977.

"Waterproofing Membranes for Bridge Deck Rehabilitation", Engineering Research and Development Bureau, New York State DOT, Albany, New York, Chamberlin, Irwin, Amsler, FHWA-NY-77-59-1, May 1977

"Corrosion of Reinforcing Steel in Concrete - A General Overview of the Problem", ASTM Special Technical Publications, Philadelphia, Pennsylvania, P.D. Cady, No. 629, 1977.

"Influence of Chloride in Reinforced Concrete", ASTM Special Technical Publications, Philadelphia, Pennsylvania, H.K. Cook, W.J. McCoy, No. 629, 1977.

"Iowa Method of Partial-Depth Portland Cement Resurfacing of Bridge Decks", ASTM Special Technical Publications, Philadelphia, Pennsylvania, E.J. O'Connor, No. 629, 1977.

"George Washington Bridge Redecked with Prefabricated Panels and No Traffic Delay", ASCE Civil Engineering, American Society of Civil Engineers, New York, N.Y., E.J. Fasullo, D.M. Hahn, Volume 47, No. 12, December 1977.

"Economic Impact of Highway Snow and Ice Control: ESIC - User's Manual", Utah DOT, Salt Lake City, Utah, FHWA-RD-77-96, 1977.

"Value Engineering Analysis of Bridge Maintenance", Wisconsin DOT, Madison, Wisconsin, Flattmeyer, December 1978.

"Repairing of Concrete Structures by Injecting Polymers - Tests of Injecting with a Column of Sand", A.M. Paillere, Y. Rizoulieres, Bull. Liaison Labo. P. et Ch., Paris, p. 17, July-August 1978.

"Orthotropic Bridge Saves Old Covered Bridge", Transportation Research Record, Transportation Research Board, Washington, D.C., R.F. Victor, No. 664, 1978.

"Stay-in-Place Forms Protect Bridge Piers", Highway and Heavy Construction, Dun-Donnelley Publishing Corporation, Chicago, Illinois, Volume 121, No. 5, May 1978.

Chapter VI

CONCLUSIONS AND RECOMMENDATIONS

VI.1 INTRODUCTION

Bridges are sections of the road network whose continual use under acceptable conditions of safety is vital for the economy of a country, but they are sensitive points of the network, particularly because of the high cost and technology involved. It is, therefore, essential to ensure serviceability at a level well adapted to traffic needs through appropriate management of the structures, and in particular, effective maintenance.

Maintenance, briefly defined as being all the operations intended to maintain the bridges in a good state of repair, should aim to:

- avoid injury to third parties or failure of a structure liable to entail tragic consequences involving responsibility of the departments in charge;
- ensure flow of traffic under the most favourable conditions possible;
- protect the national stock of bridges while striving for overall optimisation, especially from the economic point of view.

These various objectives demonstrate very clearly the interest in, and necessity for, a rational maintenance policy, especially when one considers the importance of what is at stake and the generally severe economic constraints.

VI.2 MAINTENANCE POLICY - PRESENT SITUATION

Chapter II of the report reviews the present situation of maintenance policy in a number of countries and deals with the following main aspects: objectives, current organisation, criteria to evaluate the need for maintenance, bridge replacement and maintenance costs.

The main conclusions stemming from this review are that the resources available are generally inadequate and that there are major gaps in the present state of knowledge.

In most countries, it is indicated that the funds available for bridge maintenance are insufficient and that through lack of good maintenance progressive deterioration cannot be avoided. At least 0.5 per cent of the replacement value of the bridges should be devoted yearly to maintenance expenditure in order to achieve a satisfactory standard and it is noted in some countries that there is a trend for this percentage to increase. In line with the above consideration, the value of preventive maintenance should be stressed as it avoids an accumulation of work which will place a heavy burden on future budgets. As far as resources are concerned, it should be noted that the staff work-load for both engineers and supervisors will be greater than for normal construction work as maintenance is generally widely distributed geographically and is concerned with a variety of structures.

Results of enquiries among different countries have indicated clearly a lack of knowledge in the field. This applies to maintenance costs, either globally or by bridge type, to the economic and engineering criteria to decide on maintenance works, and to the evaluation of their effectiveness. Lastly, the bridge replacement rate, which is a major factor in future planning, is very variable (0.2 - 0.4 per cent), and is based on inadequate knowledge.

Recommendations on these various aspects are given later on in this Chapter (see VI.4.1, 5.2, 6.2).

VI.3 ELABORATION OF A MAINTENANCE POLICY - GENERAL PRINCIPLES

Chapter IV of the Report "Maintenance Policy - Proposals" deals in some depth with the various problems and aspects to be considered in the elaboration of a maintenance policy and some important factors of its implementation. It may be useful to refer to it for the development of some concepts set forth hereafter.

It should be recalled that a maintenance policy constitutes an entity beginning at design stage and continuing during the whole life of the structure.

This overall aspect of the problem requires all the various elements of a management policy (inspection, maintenance, repairs, etc.) to be treated in the same spirit, i.e. the functional, economic and technical objectives must be similar and the arrangements adopted for various elements of the policy should lead to a coherent overall policy. Without proper inspection and correct evaluation of the conditions of a structure, it is impossible, for example, to achieve satisfactory maintenance and therefore to reach the overall objectives set.

Three considerations which relate to the above objectives play a part in establishing a rational maintenance policy, namely:

- safety aspects: total safety being an unattainable goal, a maintenance policy must strike a balance between the acceptable residual risks and the efforts to confine them;
- road user aspects: here, the main concern is for a level of service and comfort which should be consistent with that of the adjoining road;
- economic aspects: characterised by the fact that bridges are long-term investments where today's decisions will reach deeply into the future, as well as by technical considerations specific to individual structures.

These three groups of considerations must be integrated in a global context and taken into account with their relative weightings in different countries. The order in which these considerations are mentioned is not incidental as a conclusion of the Group's study is that safety concerns do prevail over traffic needs, which again prevail over economic and technical considerations. There is an increasing need for maintenance to be as efficient and as economic as possible in response to the present, generally difficult, economic situation. Specific mention should be made of the principal economic aspects of a maintenance policy (see IV.2.3).

Within the life span of a bridge, many decisions will be taken; those implying a choice have both an immediate and future impact. In order to rationalise the choices, it is useful to be able to compare economic data using the concept of discounting.

Discount rates are either determined beforehand or, more generally, decided according to circumstances (i.e. how much money is available). From a technical point of view, discount rates between 3 and 7 per cent seem reasonable. Discount rates for various activities and different structures should be similar but they are generally lower for major bridges located on main arteries due to traffic and economic risks incurred. Two important points are stressed: the use of discount rates implies sufficiently accurate evaluation of direct and indirect costs and the economic approach should be considered as a tool in the process of decision-making and not as an objective.

Finally, due to the complex problems to be tackled, the development of a rational maintenance policy is a difficult objective which will be aided by further studies and research in the field. Moreover, implementation of such a policy must certainly take into account all the relevant factors, even if they concern details. The main factors to be considered are the following:

- <u>Bridges</u> and their evolution.
- <u>Organisation</u> under its various aspects: responsibility for the management of maintenance at different levels, optimum planning of maintenance, criteria to be taken into account and implementation of maintenance work.
- <u>Means of action</u>, i.e. staff, finance, documentation, materials and techniques.

The proposals made in Chapter IV on this subject are repeated in the recommendations given later in paragraphs VI.4, 5 and 6.

VI.4 BRIDGES AND THEIR EVOLUTION IN TIME

It is generally acknowledged that the optimum service life of a bridge has to be long (approximately one hundred years) and that it depends mainly on the economics of maintenance as compared with strengthening or replacement.

From this viewpoint it is interesting on the one hand to take past experience into account so as to improve our knowledge and to better determine objectives and, on the other hand, to examine the dispositions and the specifications to be applied, at the different stages in the life of a structure with the aim of achieving the objectives economically.

VI.4.1 <u>Rates of bridge replacement</u>

The rates of replacement observed averaging approximately 0.2 to 0.4 per cent per annum, are rather low and will probably have a tendency to increase for various reasons set forth in Chapter II. Due to the high rate of building of new structures during the last twenty years, it is even likely that peak demands in bridge replacement will occur and these must be accommodated for. In this connection, the information available is very fragmentary and difficult to interpret correctly. It would be advisable for the various countries to carry out investigations with a view to calculating accurate replacement rates classified according to the type of structure and basic materials, and to analyse reasons for replacement.

In time, data from such enquiries will allow a better assessment of structural behaviour and a better insight of the rate of replacement.

Finally, it would be possible to analyse correctly replacement needs and to have a valuable tool for forecasting requirements.

VI.4.2 <u>Design of bridges with a view to maintenance</u>

Already at the design stage, the maintenance problem must be taken into account. This need for a good design must be particularly emphasised and it has been extensively discussed in Chapter IV.3. Some technical solutions facilitate bridge management without appreciably increasing initial costs. They are therefore highly recommended:

The main basic ideas are as follows:

- to afford easy access to the various elements of the bridge (see also report by the OECD Group on bridge inspection);
- to limit the volume of maintenance works:
  - by selection of materials (more durable concrete, special steels, low weight alloys ..);
  - by restricting aggressive agents (especially water and waste);
  - by elimination of vulnerable components (joints replaced by continuity slabs);
  - by over-design of specific components which could be very costly to repair or replace;
  - by facilitating replacement of deteriorated elements, and in particular their dismantling (grouping of vulnerable components at edges);
  - by facilitating the execution of periodic operations.

VI.4.3 Measures when opening to traffic

First, it is necessary to recall the need for the collection and compilation of the data relating to design and construction of each structure in forms to be determined by each country (see report of previous OECD Groups).

Moreover, it seems advisable to base the organisation of maintenance on instructions to be drawn up for this purpose.

The following documents should be available to the bridge management authority:

- general instructions for all bridges;
- special instructions for various kinds of bridges;
- instructions specific to each structure, taking its characteristics into account.

Instructions should give information on the periodicity of systematic maintenance operations, this periodicity being an important parameter of the economic optimisation.

VI.4.4 Measures taken during the life of the structure

The arrangements to be made are similar to those set forth above. Essentially, they are to gather and assemble all the data which is useful for the appraisal of the condition of the structure and its performance over the years (data relating to the structure, inspection reports, intervention reports, etc.).

VI.5 ORGANISATION

Organisation of maintenance must be adapted to the needs of each country as well as to those of the various managing authorities. It is, therefore, not possible to present an administrative structure suitable for universal application, but instead various principles or specific measures which are generally applicable are recommended.

VI.5.1 Responsibility for maintenance

Recommended principles of organisation are as follows (see Chapter IV.1.2):

- avoid dissociating bridges from the rest of the road network. The maintenance of bridges must be placed, except in specific cases, under the responsibility of the department in charge of the network;
- in case of intercrossing of the networks, it is generally advisable to entrust responsibility to one of the departments depending on the specific problems liable to arise or on the relative capacity of the departments in this field. It must be ensured that both departments are informed about maintenance activities in all cases. Financial problems should not interfere in any case with problems of responsibility and should be resolved at a higher level;
- performance of maintenance at various stages should be entrusted to skilled and competent staff.

VI.5.2 <u>Planning of maintenance</u>

It is advisable to proceed as follows:

- The various types of information and data relating to the evaluation of the condition of a structure should be sent to the department responsible for decisions concerning bridge management and in particular for decisions as to maintenance operations to be carried out (type and financing). This department must be fully conversant with the facts so as to be able to interpret the information and reach appropriate decisions to ensure safety of traffic. Reciprocally, decisions reached must be suitable for application without any ambiguity, at the technical level. This arrangement demonstrates the <u>necessity of a common language</u> for implementation of rational and uniform maintenance policies. Moreover, different administrative arrangements can improve the efficiency of the system, e.g. extension of regulations to all bridges in a country, existence of a central department entrusted in particular with the formulation of technical standards and general strategy.
- Decisions reached should tend towards an <u>overall optimisation from the economic point of view</u>. In this respect, two main aspects should be taken into consideration. On the one hand, it is accepted that maintenance must have a preventive character in order to avoid any deterioration, the repair of which would be more costly than preventive operations. It is the aim of good management to minimise costs during the life of the bridges. On the other hand, there are often constraints of a financial kind implying choice, either between different activities, or within the very management policy itself. As regards the economics involved, it is essential that these choices be made in a rational manner (see chapter IV.3) and based on precise data. The optimum solution therefore requires complicated studies and important statistical information not generally at our disposal at present. Therefore, it would be advisable to gather information to be used as a basis for rationalisation of maintenance operations as regards the technical and economic aspects involved and to process most of this information in a data bank for reasons of efficiency. Among these data must be included the real costs and details of maintenance given as a function of various factors (type and age of the structure, traffic, etc.), the problems arising frequently, the type of maintenance which is particularly expensive and its cause, the results of maintenance carried out and discount rates applied.

VI.5.3 Conduct of maintenance operations

Two categories of maintenance should be differentiated:

- <u>ordinary maintenance</u> requiring few resources and little technical capacity. These maintenance operations should generally be carried out by staff belonging to the department responsible for regular maintenance and should be done in close co-operation with the inspection team;
- <u>specialised maintenance</u> relating to operations requiring either techniques to be applied by specialists or the use of special resources. These maintenance operations should generally be carried out by specialised enterprises having the necessary equipment. It would probably be advisable to seek a contractor skilled in maintenance of bridges.

In these two major categories, the work of maintenance can be either planned in advance and carried out periodically or, on the contrary, it may not be foreseeable because it results from observations made during the course of inspection.

VI.6 MEANS OF ACTION

VI.6.1 Personnel

The maintenance of structures (as indicated in VI.2) requires more staff, whether engineers or technicians, than for new construction.

To attain efficient and economic maintenance, it is necessary that the maintenance measures are decided upon and applied in good time and that the techniques are correctly used. For this to be done, account must be taken, on the one hand, of the particular characteristics of the structures and of the techniques, on the other, the multi-disciplinary nature of maintenance works requires professional engineers with a broad knowledge as well as specialists. It is, therefore, essential that maintenance teams are staffed with able and skilled personnel.

Since competence results either from specialisation or, above all, from individual experience, it is desirable to have a low turnover of personnel.

Efforts should be devoted to training centred on the following:

- ensuring that the importance of maintenance is fully recognised at all levels concerned;
- ensuring application of uniform maintenance methods and techniques by competent staff.

Training methods are very diversified. It should be noted that, owing to the relatively recent importance given to maintenance, the development of training schemes would make it necessary to increase the exchange of information in this field not only between maintenance personnel but also on an international level.

VI.6.2 Level of expenditure

An annual expenditure at least equal to 0.5 per cent of the replacement cost of the bridges seems to be necessary to implement a rational policy of preventive maintenance. Some factors set forth in Chapter II seem to imply that this level of resources is insufficient.

Replacement cost varies not only with bridge types but also with the nature of replacement works (i.e., replacement of deck using the same supports), and local conditions (disturbance to traffic if the new bridge remains at the same location). This cost can then be very different from the construction cost. Present knowledge on the replacement value of bridges is very insufficient and it would be useful to implement accounting systems for bridge maintenance and, more generally, for bridge management, in order to estimate more precisely the level of resources needed.

VI.6.3 Documentation - data bank

Recommendations on this subject, which are quite valid as far as maintenance is concerned, have already been made in reports by preceding OECD groups relating to inspection and load carrying capacity of bridges as well as in chapters of this report.

There is a strong inter-action between maintenance and bridge inspection. The work to be done and its planning are based primarily on results of bridge inspection; the results obtained from maintenance work already carried out should be checked by inspection.

In short, necessary documentation consists of two main parts meeting two aims:

- <u>Knowledge of the subject</u> (see report on bridge inspection) with systematic collection of all data relating to bridges, their behaviour, and their rational use at local and central levels. In this respect it seems advisable to promote a methodology of data processing and thus the use of data banks. Appendix C to Chapter IV is a good example of what could be done. Nevertheless, proper consideration should be given to the precautions needed in the operation of data banks (see IV.3.), and especially as concerns:
  . the limited amount of information that can be introduced and the selection of data so as to facilitate the regular updating of the system;
  . the need for a very thorough training of the personnel operating the data bank.
- <u>Use of a common language and methods</u> (at least within each country). In this respect various kinds of documents should be provided for and drawn up:
  . general and special instruction booklets classified by type of bridges;
  . lists of parts of structures and catalogues of defects. These documents which at the same time must be simple, correct and functional, are necessary for making a good diagnosis, which will be clearly understood by all parties. An example of such a catalogue is given in Appendix A to this report. Quality "standards" determining thresholds at which corrective action needs to be taken should be added to these catalogues;
  . guides relating to various maintenance techniques. To correct a given defect, there may be several techniques and a given technique may sometimes correct several defects. These guides are, therefore, particularly useful for selecting an appropriate technique and implementing it.

VI.6.4 Maintenance equipment and techniques

Some maintenance techniques and the equipment used have been described in Chapter III. This description presents a review of our present knowledge in this field, but cannot be comprehensive owing to the variety of techniques used.

Moreover, the development of methods for the maintenance of structures is relatively recent; improvements in present techniques are bound to be made and new techniques will certainly be developed in coming years. It should also be noted that the rapid technological evolution in the design and construction of bridges as well as the increasing desire to use very sophisticated techniques and equipment will surely favour these developments.

There is, therefore, a research need, especially in areas where present techniques are not very successful (i.e., waterproofing, buried joints, materials for repairing concrete ..) and also concerning long-term behaviour of some synthetic materials and development of new techniques. Detailed recommendations on this subject are given in Chapter V.

VI.7 GENERAL CONCLUSION

The elaboration and implementation of an as rational as possible maintenance policy is a major challenge which has only quite recently come to the forefront. The need to obtain more detailed and extensive knowledge in the field of bridge maintenance exists in many countries.

Furthermore, owing to the international nature of traffic, problems that arise, even if they are not identical from one country to the next, are sufficiently similar and of such importance as to justify the necessity of international co-operation. This co-operation is advisable not only at the level of research work to be undertaken and techniques to be implemented but also in regard to methodology, decision-making criteria and experience gained.

Annex A

EXTRACTS FROM: "DEFAUTS APPARENTS DES OUVRAGES D'ART EN BETON" (1)

This publication, prepared by the Service d'Etudes Techniques des Routes et Autoroutes and the Laboratoire Central des Ponts et Chaussées, consists of:

Presentation,
Introduction,
Table recapitulating defects classified according to their index of severity,
Catalogue of defects in the structure,
Defects in the bearings,
Defects in joints, at the road surfacing,
Defects in the road surfacing.

The introduction and extracts from the catalogue of defects are reproduced below. Reference is made to the publication "Inspection, maintenance and repair of structures"(2) prepared by the Service d'Etudes Techniques des Routes et Autoroutes.

INTRODUCTION

This catalogue of defects in concrete structures is published for the use of staff responsible for the surveillance and inspection of bridges, such as is advocated in the document SERO 70. Its aim is to secure greater uniformity in the viewpoint of these staff, which would seem to be essential.

This catalogue takes account of five kinds of defect, each being assigned an index letter.

B - Defects present from the time of construction and without important consequences apart from aesthetics.

C - Defects which indicate the risk of abnormal developments.
    - Defects revealing the development of deterioration of the structure. They are placed in two classes:

DA - Defects which indicate the first appearance of a development.

DB - Defects which indicate an advanced development.

E - Defects which show, in a very distinct manner, a change in the behaviour of the structure and which bring into question the duration of life of the structure.

F - Defects indicating the approach to a limit state and necessitating a restriction on use, rendering the structure unserviceable.

---

1) Défauts apparents des ouvrages d'art en béton. Paris, 1975. Extracts by permission of the Ministère de l'Equipement : Service d'Etudes Techniques des Routes et Autoroutes and of the Laboratoire Central des Ponts et Chaussées.

2) Surveillance, Entretien, Réparation des Ouvrages d'Art (SERO 70). Ministère de l'Equipement : Service d'Etudes Techniques des Routes et Autoroutes, Paris, 1977.

The defects are classified in alphabetical order. For each defect the following is given:

A definition.
Some probable causes for its appearance.
In most cases a photographic illustration and/or a sketch.
The index letter defining the severity of the defect, this index also makes it easier to prepare a computerised index of structures.

Name of Defect	Index of Severity	Definition – Probable Cause	Illustration	Observations or Sketch
BLISTERS		See geometrical defects		
BLEEDING CHANNELS OF CONCRETE, ETC.		See geometrical defects		
CARBONATION	B	Transformation of lime into carbonate by the action of carbonic gas. It shows itself in two forms: 1) Formation of a surface deposit under the effect of the surrounding atmosphere (flow of water, environment) – On load carrying elements 2) Appearance of white marks (see efflorescence)		
CRACKS (FRACTURES)	E F	Very important cracks extending right through the concrete, they are often accompanied by spalling and/or deformation (see these defects) – On non-load carrying element – On load carrying element		

112

Name of Defect	Index of Severity	Definition – Probable Cause	Illustration	Observations or Sketch
EFFLORESCENCE		White stains on the surface of concrete arising from carbonation (see this defect) Probable cause: bad waterproofing of the structure		DA — Extensive efflorescence under cantilever
	DA	In the case of part of a reinforced concrete structure, a significant number (of defects) on non-load carrying elements or isolated (defects) on load-carrying elements		
	DB	In the case of reinforced concrete, a significant number (of defects) or concentrated (defects) on a load-carrying element In the case of prestressed concrete (defects) located at places related to the position of tendons		DB — Efflorescence on beam Efflorescence along a prestressing tendon

113

Name of Defect	Index of Severity	Definition - Probable Cause	Illustration	Observations or Sketch
CRACK ALONG THE PRESTRESS-ING CABLES (cont.)		- Cracks long the ducts which develop to more than 0.3 mm		
LONGITUDINAL CRACK (vertical in the case of piers, walls or abutments)		- Cracks parallel to the longitudinal axis of the structure. State the length of the crack, its width, and the resulting spacing in the case of several cracks		Piers View underneath the bridge deck e = width of crack l = length of crack
	C	- Cracks appearing at construction in parts of a reinforced concrete sturcture, not affecting the stability of the structure and with e < 0.5 mm. In the case of walls this limit may be taken to 3 mm.		
	DA	- For reinforced concrete openings of less than 0.5 mm since construction or the appearance of the cracks.		

Name of Defect	Index of Severity	Definition – Probable Cause	Illustration	Observations or Sketch
LONGITUDINAL CRACKS (cont'd.)	DB	- For reinforced concrete widths between 0.5 mm and 2 mm. - For prestressed concrete width < 0.3 mm.		In reinforced concrete
	E	- In the case of reinforced concrete cracks which develop to more than 2 mm.		
	F	- In the case of prestressed concrete opening of more than 0.3 mm. - Cracks of which the location and development endangers the use of the structure		

Annex B

DATA ON THE REPLACEMENT OF BRIDGES IN FRANCE DURING 1978

The number of bridges replaced in 1978 was 141 and their total surface area 31,600 m^2 (average 224 m^2 per bridge). In 1976 and 1977 the numbers replaced were 137 and 148 respectively. The number of bridges in France is 50,000, thus giving an annual rate of replacement of just under 0.3 per cent for the three years.

Particulars of the bridges replaced in 1978 are:

Construction material:

masonry	27%
reinforced concrete	13%
metal	30%
no information	30%

Reason for replacement:

change of geometry	37%
change of load-carrying capacity	13%
degradation	28%
various or no information	22%

Annex C

EXAMPLE OF A BRIDGE DATA BANK SYSTEM

1. INTRODUCTION

The Road Directorate under the Danish Ministry of Transport is the responsible authority for motorways and trunk roads in Denmark. Methodical inspections of structures on the motorways were implemented in 1971, and this exercise was in 1972 extended to the trunk roads. Most of the structures are bridges, but retaining walls, ferry berths, etc. are also included.

Inspection reports are available for all structures covered by the system. The reports are used, i.e. to administer the maintenance needs detected.

The set of reports, furthermore, contains valuable information of a more general nature, about the frequency of problems of certain types, and on which structural parts they occur. When this information is common knowledge, it can become the basis for attempts to reduce or prevent the appearance of such problems. It requires, however, considerable efforts to process the data manually in a useful form.

The application of computer techniques reduces such work, and the Road Directorate has, therefore, investigated this possibility. As a start a basis was formulated. This was done by "Basic Views" dated November 1976. Based on this paper a Working Committee was established in 1977, with the terms of reference to prepare an advance study report for the establishment of a computer based <u>Register of Bridge Inspections</u>.

In "Basic Views" it had been emphasised that computer processing of the data in the inspection reports would only give limited benefits compared to the benefits obtainable if it was possible at the same time to have access to the general technical and administrative information about the structures contained in a computer based <u>Register of Bridge Data</u>.

Already earlier the establishment of a General Register of Road Data had been started by the Road Directorate, which included a section for structures, mainly to be used for selection of routes for special road transports. At the same time, the Bridge Division was working on a manual register of structures, containing administrative, economic, geometrical and structural information.

It was, therefore, decided that the Working Committee in the advance study report should also analyse the consequences of the establishment of a Register of Bridge Data covering the above three demands, and make recommendations for its framework.

In preparing the recommendation, co-ordination should be made with the draft Manual for Inspection of Structures prepared by the Committee for Road Standards. This draft has later been approved.

The "Advance Study Report", published 1978-05-18, contains an account of the background and includes the following:

<u>Existing operation procedure for data about structures</u>

The present manual way of operating the archives of the Bridge Division in the Road Directorate is described. The operation lies partly with the <u>Bridge Archive</u>, responsible for the List of Bridges, the List of Drawings and Administrative Archives, partly with the Maintenance Section, which conducts the inspection of structures and prepares the inspection reports.

The benefits achieved by introducing computer processing of incoming data and of the updating of a Register of Bridge Data and Bridge Inspections are described, it is pointed out which rationalisation such system will give and which new tasks can be solved. Typical examples of analysis, which will be possible by computer processing, are shown.

<u>Draft register of bridge data and draft register of bridge inspections</u>

The logical structure of the Registers is shown, and it is described how the connection between the two Registers can be established. The relation to the Register of Road Data is described, as well as it is shown that the data in both the Bridge and the Road Registers can be used simultaneously.

A summary is given of the data contained in the two registers, and the operation and updating of the Registers of Bridges is described.

It will be possible to use the system developed for the General Register of Roads by the two Registers of Bridges.

<u>Use of Registers</u>

The section includes some general considerations about the use of the Registers of Bridges regarding planning, operation and maintenance of bridges.

Further is described the basis-reports, the special reports and other uses which can be made by the data contained in the registers.

Consequences

The consequences of the establishment of Registers of Bridges are explained both in the building-up phase and when in operation.

It is assumed that the system shall be operated by the existing personnel, who should be able - due to the rationalisation through the computer processing - to conduct the analyses, which have not been possible before with the time-consuming manual system. Additional resources will, however, be required in the register of Roads Section for the computer operation.

Based on the above, the Road Directorate has decided to commence the work on establishing a Register of Bridge Data and a Register of Bridge Inspections.

The Laboratory of Road Data Processing, the computer division under the Road Directorate, has prepared a "User's Manual", published as a loose-leaf book.

2. REGISTERS OF BRIDGES, BASIC STRUCTURE

The structure comprises in principle two different registers:

- Register of Bridge Data
- Register of Bridge Inspections

The connection between the two Registers is ensured through the common identification of structures, which is the key-input to all other data in the two registers.

The Register of Bridge Data contains two groups of data:

- Administrative and technical/economical data for the structure itself, and
- Pasage information (data about the passing traffic routes, including rating of the bridge).

The Register of Bridge Inspections contains also two groups of data:

- Data for a historical summary, and
- Information about problems which is split into:
    - Problem location information
    - Problem description information

The logical build-up of the Registers is shown in the following diagram:

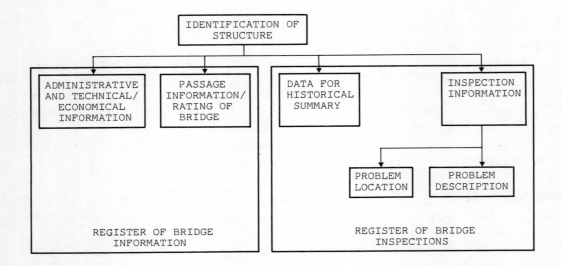

Identification of Structures

The connection between a structure and all related information is achieved through

BYGVAERKSIDENTIFIKATION (Identification of structures)

The identification is a 12-digit number constructed as follows:

```
AAARRRRPNNNN
```

AAA	Authority (owner of structure)
RRRR	Road No.
P	Part of road (if separate track for each carriageway)
NNNN	Serial No.

The first part of the identification (AAARRRRP) has the same form as the Identification of Roads used by the General Register of Roads, and, therefore, a joint run is possible.

Administrative Information

The administrative description comprises the elements:

BYGVAERKSBETEGNELSE	(Designation of structure)
UTM-REFERENCE	(Reference to UTM-System)
DSB BRO NR	(Railway bridge no.)
OPFØRT AR	(Erection year)
AENDRET AR	(Year of major changes)

These elements, together with the Identification of Structure, constitute the List of Structures.

Technical/Economical Information

The technical/economical description comprises the elements:

BYGVAERKSTYPE	(Type of structure)
BROLAENGDE	(Length of bridge)
BROBREDDE	(Width of bridge)
SKAERINGSVINKEL	(Angle of intersection)
SAMLEDE UDGIFTER	(Total cost)
SAMARBEJDSPART	(Participating authority)
EFTERSYNSANSVARLIG	(Responsible for inspection)
PROJEKTERENDE	(Consultants)
FUNDERING	(Foundation type)
SPAENVIDDER	(Bridge spans)
KODE FOR SVAG BRO/SKILT	(Code for bridge with limited traffic load/traffic sign)
BESKRIVELSE AF SKILTNING VED SVAG BRO (TEKST)	(Description of traffic sign by bridge with limited traffic load (text))

The above data have been selected among many. The scope of possible analyses is very much dependent on the amount of data, but a limitation has been necessary.

Information about Passages

A passage is a traffic route passing a structure as an overpass, an underpass or possibly a bypass.

"Traffic Route" and "Structure" have a very wide definition as shown below:

Examples:

1. A highway passes under a bridge which carries a minor road across the highway. The bridge, therefore, has two passages:
   - the highway as an underpass passage;
   - the minor road as an overpass passage.
2. A bridge is carrying a highway across a stream. Thus, there are two passages:
   - the highway as an overpass passage;
   - the stream as an underpass passage.
3. A bridge carries a highway across an express road. There are then three passages:
   - the one carriageway of the express road as an underpass passage;
   - the other carriageway of the express road as an underpass passage;
   - the highway as an overpass passage.

The examples show that several passages can be connected with a structure. To identify the different passages, each passage is given a number.

PASSAGENUMMER (Identification No. of passage)

which for each structure can be from 1 to 99. Each of the numbered passages can be described by a set of passage information, which comprises the following elements:

PASSAGE TYPE	(Type of passage)
PASSAGE O/U/A	(Locate the passage in relation to the Structure)
VEJIDENTIFIKATION	(Identification of the road)
KILOMETERING	(Stationing)
FRITRUMSPROFIL	(Clearance)
BAEREEVNEBEDØMMELSE (GAMMEL DEL/NY DEL)	(Evaluation of load carrying capacity, Old part/new part)
BEMAERKNINGER TIL BAEREEVNEBEDØMMELSE (TEKST)	(Remarks to evaluation of load carrying capacity (text))
BAEREEVNEKLASSIFI-KATION	(Rating of Bridge)
PLACERING AF SPECIALSPOR (TEKST)	(Location of specially-rated lanes (text))

Not all elements will be relevant for a particular passage, for example, will it not be necessary to record the load carrying capacity for the underpass passage (highway) in Example No. 1 above?

Historical Summary

To the historical summary is reported remarks, which are found necessary to describe the history of the structure. The historical summary is made by collecting all these remarks and listing them in the sequence they have been reported.

After each inspection of a structure, its condition is to be described by the following elements:

VEJRFORHOLD	(Weather condition at inspection)
TILSTANDSKARAKTER	(Rating of the conditions of structure)
BEMAERKNING TIL KRONOLOGISK OVERSIGT (TEKST)	(Remarks to Historical Summary (Text))

Further can be reported:

EFTERSYNSINTERVAL	(Inspection interval)
VEDLIGEHOLDELSES-OMKOSTNINGER	(Maintenance costs)

Location of Problem

In order to enable reference to different locations of problems, when they have been reported, each location is given a

    SKADENUMMER   (Problem Identification No.)

which for each structure can be from 1 to 999. The numbered location of problem is described by the following elements:

KONSTRUKTIONSELEMENT	(Structural part and type)
ELEMENTPLACERING	(Location of structural part)
DEL AF ELEMENT	(Section of the structural part)
DETAILPLACERING	(Location in detail)

which will make it possible to state the problem's exact location at the structure.

Description of Problem

The description of a problem consists of the following elements:

SKADEBESKRIVELSE	(Description of problem)
FOTONUMMER	(Photo Identification Number)
UDBEDRING	(Recommendation for correction)
RAPPORT	(Report made)

The elements make it possible to give a detailed description of a problem as it is found at an inspection. Further, it is possible to note, if the problem should be corrected and how, and finally it is possible to note if the requested correction has been done, and the date for such correction.

The description of the problem is very detailed and it may be necessary to attach several descriptions of problem to the same location of problem (Problem Identification No.). In order to be able to distinguish between these descriptions of problem, each one shall be given a letter as identification.

3. REGISTERS OF BRIDGES, OPERATION AND UPDATING

An important point at the construction of administrative registers is the establishment of the organisation which is going to ensure the daily operation and updating of the collected information.

The selected organisation consists of a central section at the Laboratory of Road Data Processing, which will conduct the daily operation of the registers and computer programmes, and of key personnel at the Bridge Division's Bridge Archive and in the Maintenance Section, who will be responsible for the daily reporting and the use of information (new recording and alterations). The central section advises the key personnel on doubtful questions and receives criticism and recommendations regarding alterations and improvements to the registers.

4. REGISTERS OF BRIDGES, USE

The applications can be shown graphically:

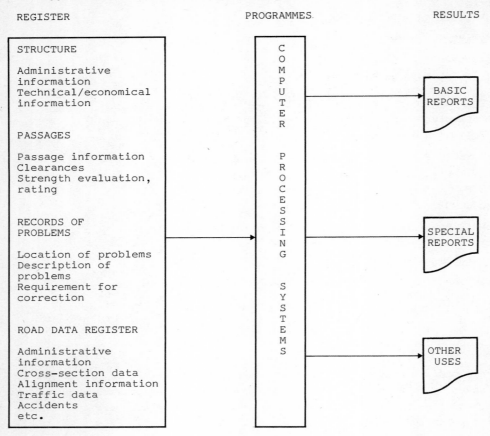

4.1 Basic Reports

The basic reports are the core of the computer processing registers of bridges. They have several functions:

- They act as receipts to the user for data reported to the system.
- They act as local manual files for bridges and bridge problems.
- The basic reports, or parts thereof, can be ordered and used for actual analyses.

There are four different basic reports in the system:

- The Structures
- The Passages
- The List of Problems
- The Historical Summary

The basic reports contain the same data elements as the corresponding input-forms and are printed automatically whenever information has been reported.

Further, basic reports for each structure can be ordered independently.

4.2 Special reports and other uses

Special computer programmes are developed to utilise the information in the Registers of Bridges in Special Reports for a number of purposes.

Furthermore, a computer programme able to carry out different types of analyses of the data content in the Registers based on user-decided questions is at present being developed.

Presently, requests have been made for the following special reports and types of analyses:

List of Structures

The list of structures in the Road Directorate can easily be updated by using the information contained in the computer register of bridge data. The list primarily contains the administrative information supplemented with the identification of structure.

Preparation of lists of structures with special characteristics

The Bridge Archive of the Bridge Division will be able to request summaries of structures which have a particular characteristic or a special combination of characteristics, for instance, questions of types below:

- Which bridges are prestressed with cables, and how is the distribution of these bridges regarding the year of construction?
- Which bridges have wooden piles?
- Which bridges are substandard regarding load carrying capacity?

Administration in the Bridge Division's Maintenance Section

As a help for the Maintenance Section in the daily work can be mentioned reports and results on different matters, which can be made by the computer system of registers.

- When scheduling inspection of structures, the inspection officer can request a list of those structures which are to be inspected the said year.
- When establishing priorities for maintenance work, the officer in charge can request a list of structures, where the structures are listed after their rating of condition, which, in principle, is the same as a list of priority.
- If special maintenance work is considered to be done on a series of bridges, a list of priority can be printed out for these particular types of maintenance work. As an example can be mentioned painting of guardrails, cleaning of joints, and minor patching-up works.
- Regarding management, questions of the types below can easily be answered:
  - Which bridges have certain problems not yet scheduled for repair?
  - For which structures have repair works been ordered, but not reported to be completed?

Analysis of Problems

The primary object of establishing the computerised registers of bridges is to make it possible to analyse all problems, taking into account the total amount of bridges and other structures. The work of analysis is necessary to utilise the knowledge of problems collected at the inspections, and thereby preventing similar problems at future structures.

As an example can be mentioned:

- Which (how many) bridges have reduced concrete cover on prestressed beams?
- Which (how many) bridges are constructed with prestressed concrete beams?

- Which bridges have cracks of the pavement at the abutments?
- Which bridges have the same length?

- Which bridges have map cracking?
- Towards which quarter?

- Which problems are the most common?
- How often?

The content of data in the computerised register of bridges is available for research work. The majority of all needs can be handled by the basic reports, the special reports, or the computer programme of questions. Only in exceptional cases it will be necessary to develop new special computer programmes.

Annex D

EXAMPLE ILLUSTRATING CONSEQUENCES OF VARYING DISCOUNT RATES

Let us assume that we have just finished the construction of a bascule bridge at a total cost of 100 mio Dkr (see below). We further assume knowledge of the future costs such as annual operation and maintenance costs of Dkr 2 M, major repairs after 10, 30 and 50 years as indicated in Figure D1. After 70 years we assume a total replacement costing Dkr 150 M. All amounts at present price level.

A total life cycle lasting 70 years and costing Dkr 350 M without discounting the costs.

COSTS OF A LIFE CYCLE
Initial construction costs: Dkr 100 M
FUTURE COSTS:
Annual costs for operation and maintenance: 2 M
    Repairs after 10 years: 10 M in one year
    "   "   30   "   :   20 M in two years
    "   "   50   "   :   30 M in two years
Replacement after 70 years :   150 M in three years

TOTAL COSTS OF A LIFE CYCLE AFTER INITIAL
CONSTRUCTION WITHOUT DISCOUNTING:

Operation and maintenance: 140 M
Repairs: 60 M
Replacement: 150 M

Total: 350 M

This is an example of a discount rate = 0%.
Let us consider a discount rate = 4%.

Figure C 1 graphically illustrates the present value of the future costs = 350 M, when the discount rate is 4 per cent. To aid the illustration the abscissa (time) has been discounted instead of the ordinate (future costs). The total area expresses the total present value of the total future costs.

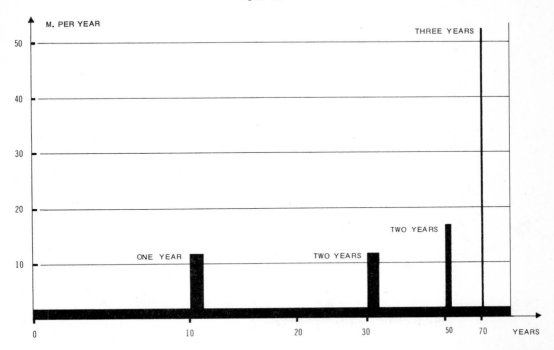

Figure D.1

It can be observed that the repair after 10 years costing 10 M has almost the same present value as the replacement after 70 years costing 150 M.

The total costs with a discount rate = 7 per cent are shown on figure D 2.

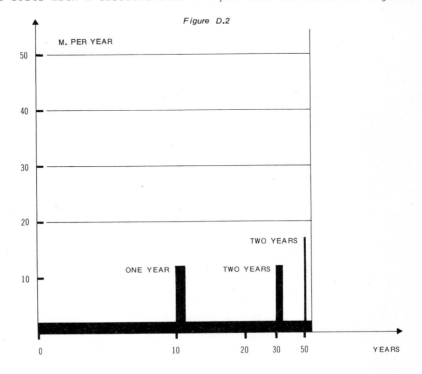

Figure D.2

With 7 per cent the replacement of the bridge has practically lost its present value and so have the repairs after 50 years at Dkr 30 M. The economic horizon does not stretch much further than 30 years.

Figure D 3 illustrates the situation with a discount rate of 10 per cent.

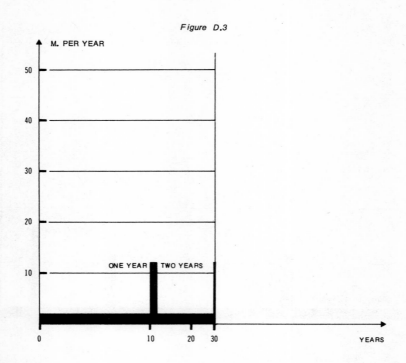

Figure D.3

With 10 per cent the economic horizon is limited to about 20 years.

Table D.1 gives the figures for the four discount rates. It should be noted that the annual operation and maintenance costs play a very significant role regardless of the discount rate.

Table D.1

Discount Rates	0%	4%	7%	10%
	M	M	M	M
Operation and maintenance in 70 years - 2 M annually	140	47	28	20
Repairs after 10 years - 10 M in one year	10	7	5	4
Repairs after 30 years - 20 M in two years	20	6	2	1
Repairs after 50 years - 30 M in two years	30	4	1	0
Reconstruction after 70 years - 150 M in three years	150	9	1	0
Present values of total discounted costs in 70 years	350	73	37	25

As this example illustrates, the discount rate is decisive when considering the level of quality (durability) that should govern our design and construction of bridges as well as the zeal with which maintenance should be carried out.

LIST OF MEMBERS OF THE GROUP

Chairman: Mr. M. Le Franc

BELGIUM
    Mr. L. MAHIEU
    Inspecteur général ff. des Ponts et Chaussées
    Ministère des Travaux Publics
    Bureau des Ponts - 1ère Division
    Rue Guimard 9
    B-1040 BRUXELLES

    Mr. V. VEVERKA
    Ingénieur civil
    Chef du Service Conception et Structure
    Centre de Recherche Routière
    Boulevard de la Woluwe 42
    B-1200 BRUXELLES

CANADA
    Mr. P. CSAGOLY
    Head, Structural Research
    Ontario Ministry of Transportation
     and Communications
    1201 Wilson Avenue
    DOWNSVIEW, Ont. M3M 1J8

DENMARK
    Mr. H.H. GOTFREDSEN, M.Sc.
    Danish Ministry of Transport
    The Road Directorate
    Postbox 2169
    Havnegade 23
    DK-1016 COPENHAGEN K

FINLAND
    Mr. Kari MOIJANEN, Civ. Eng.
    Roads and Waterways Administration
    Opastinsilta 11-12
    00520 HELSINKI 52

FRANCE
    Mr. C. BOIS
    Chef du Département des Structures et
     Ouvrages d'Art
    L.C.P.C.
    58 Boulevard Lefebvre
    75732 PARIS CEDEX 15

    Mr. M. LE FRANC
    I.C.P.C.
    Conseil Général des Ponts et Chaussées
    246 Boulevard Saint-Germain
    75775 PARIS CEDEX 16

    Mr. Ph. LEGER
    Direction régionale de l'Equipement
     de l'Ile de France
    23 rue Miollis
    75732 PARIS CEDEX 15

GERMANY
    Regierungsdirektor D.E. LEBEK
    Bundesanstalt für Strassenwesen
    Brühler Strasse 1
    D-5000 KOLN

ITALY	Ing. Gabriele CAMOMILLA Società Autostrade Via Nibby 10 I-00160 ROME
	Ing. Emanuele SCOTTO Direzione Generale ANAS Via Monzambano 10 I-00185 ROME
JAPAN	Mr. Shoich Saeki Chief, Bridge Division Public Works Research Institute Ministry of Construction Asahi 1-branch, Toyosato-chyo TSUKUBA-GUN, IBARAGI-KEN
NETHERLANDS	Ir. H. de VRIEND Directie Bruggen van de Rijkswaterstaat Postbus 285 2270 AG VOORBURG
	Ir. R. VELLEMA Directie Bruggen van de Rijkswaterstaat Postbus 285 2270 AG VOORBURG
NORWAY	Mr. Knut NAESS Chief Engineer Norwegian Public Roads Administration Bridge Division P.O. Box 8109 OSLO-DEP
SPAIN	Mr. Ramon Del CUVILLO Jefe de la Seccion de Estructuras Servicio de Tecnologia de la Carretera Direccion General de Carreteras Ministerio de Obras Publicas y Urbanismo MADRID 3
	Mr. Amadeo Trias GONZALEZ Ingeniero Jefe del Servicio de Conservacion Direccion General de Carreteras Ministerio de Obras Publicas y Urbanismo MADRID 3
SWEDEN	Mr. Hilding PERSSON National Swedish Road Administration Fack S-102 20 STOCKHOLM 12
SWITZERLAND	Mr. Edmond REY Adjoint scientifique au Service fédéral des Routes et des Digues Monbijoustrasse 40 CH-3003 BERNE
UNITED KINGDOM	Mr. J.A. LOE Structures Department Transport and Road Research Laboratory Old Wokingham Road CROWTHORNE, Berks. RG11 6AU
	Mr. G.P. MALLETT Department of Transport St. Christopher House Southwark Street LONDON SE1 OTE

UNITED STATES   Mr. Stanley S. GORDON
                                Bridge Division
                                Federal Highway Administration
                                WASHINGTON D.C. 20590

                                Mr. Charles F. GALAMBOS
                                Structures and Applied Mechanics Division
                                Office of Research
                                Federal Highway Administration
                                Department of Transportation
                                WASHINGTON D.C. 20590

                                Mr. Donald A. LINGER
                                Chief, Bridge Structures Group
                                Office of Research
                                Federal Highway Administration
                                Department of Transportation
                                WASHINGTON D.C. 20590

OECD SECRETARIAT   Mr. B. HORN
                                Mr. B. HUEBER
                                Mme. E. FRUTON

Rapporteurs   Messrs. CAMOMILLA, GALAMBOS, GOTFREDSEN, LE FRANC,
                                        LINGER, LOE

Members of the Editing   Messrs. CAMOMILLA, GORDON, GOTFREDSEN, LEBEK, LE FRANC,
Committee   LOE, MAHIEU, MALLETT, HORN, HUEBER

LIST OF ROAD RESEARCH PUBLICATIONS

Road Construction and Maintenance

Research on crash barriers (February 1969)
Motor vehicle corrosion and influence of de-icing chemicals (October 1969)
Winter damage to road pavements (May 1972)
Accelerated methods of life-testing pavements (May 1972)
Proceedings of the symposium on the quality of road works (July 1972)
Waterproofing of concrete bridge decks (July 1972)
Optimisation of road alignment by the use of computers (July 1973)
Water in roads: prediction of moisture content in road subgrades (August 1973)
Maintenance of rural roads (August 1973)
Water in roads: methods for determining soil moisture and pore water tension (December 1973)
Proceedings of the symposium on frost action on roads (October 1974)
Road markings and delineation (February 1975)
Resistance of flexible pavements to plastic deformatinn (June 1975)
Bridge inspection (July 1976)
Road strengthening (September 1976)
Proceedings of the symposium on road design standards (May 1977)
Use of waste materials and by-products in road construction (September 1977)
Maintenance techniques for road surfacings (October 1978)
Catalogue of road surface deficiencies (October 1978)
Evaluation of load carrying capacity of bridges (December 1979)
Construction of roads on compressible soils (December 1979)
Proceedings of the symposium on road drainage (January 1980)(*)

Road Transport and Urban Transport

Electronic aids for freeway operation (April 1971)
Area traffic control systems (February 1972)
Optimisation of bus operation in urban areas (May 1972)
Two-lane rural roads: road design and traffic flow (July 1972)
Traffic operation at sites of temporary obstruction (February 1973)
Effects of traffic and roads on the environment in urban areas (July 1973)
Proceedings of the symposium on techniques of improving urban conditions by restraint of
   road traffic (September 1973)
Urban traffic models: possibilities for simplification (August 1974)
Capacity of at-grade junctions (November 1974)
Proceedings of the symposium on roads and the urban environment (October 1975)
Research on traffic corridor control (November 1975)
Bus lanes and busway systems (December 1976)
Transport requirements for urban communities (December 1977)
Energy problems and urban and suburban transport (December 1977)
Integrated urban traffic management (December 1977)
Traffic measurement methods for urban and suburban areas (March 1979)
Transport services in low density areas (September 1979)
Management of urban freight distribution (derestricted in October 1980)(*)
Urban Public Transport: Evaluation of Performance (October 1980)
Transport choices for urban passengers: measures and models (September 1980)
Evaluation of urban parking systems (December 1980)

Road Safety

Alcohol and drugs (January 1968)
Pedestrian safety (October 1969)
Driver behaviour (June 1970)
Proceedings of the symposium on the use of statistical methods in the analysis of road
   accidents (September 1970)
Lighting, visibility and accidents (March 1971)

---

*) Can be obtained on special request from OECD Road Research Secretariat

Research into road safety at junctions in urban areas (October 1971)
Road safety campaigns: design and evaluation (December 1971)
Speed limits outside built-up areas (August 1972)
Research on traffic law enforcement (April 1974)
Young driver accidents (March 1975)
Roadside obstacles (August 1975)
Manual on road safety campaigns (September 1975)
Polarised light for vehicle headlamps (December 1975)
Driver instruction (March 1976)
Adverse weather, reduced visibility and road safety (August 1976)
Hazardous road locations: identification and countermeasures (September 1976)
Safety of two-wheelers (March 1978)
New research on the role of alcohol and drugs in road accidents (September 1978)
Traffic safety in residential areas (October 1979)
Road safety at night (December 1979)
Summary record of the Symposium on Safety of Pedestrians and Cyclists (derestricted in May 1980)(*)
Guidelines for driver instruction (March 1981)
Traffic control in saturated conditions (January 1981)
Methods for evaluating road safety measures (June 1981)

---

*) Can be obtained on special request from OECD Road Research Secretariat.

OECD SALES AGENTS
DÉPOSITAIRES DES PUBLICATIONS DE L'OCDE

ARGENTINA – ARGENTINE
Carlos Hirsch S.R.L., Florida 165, 4° Piso (Galería Guemes)
1333 BUENOS AIRES, Tel. 33.1787.2391 y 30.7122

AUSTRALIA – AUSTRALIE
Australia and New Zealand Book Company Pty, Ltd.,
10 Aquatic Drive, Frenchs Forest, N.S.W. 2086
P.O. Box 459, BROOKVALE, N.S.W. 2100

AUSTRIA – AUTRICHE
OECD Publications and Information Center
4 Simrockstrasse 5300 BONN. Tel. (0228) 21.60.45
Local Agent/Agent local :
Gerold and Co., Graben 31, WIEN 1. Tel. 52.22.35

BELGIUM – BELGIQUE
LCLS
35, avenue de Stalingrad, 1000 BRUXELLES. Tel. 02.512.89.74

BRAZIL – BRÉSIL
Mestre Jou S.A., Rua Guaipa 518,
Caixa Postal 24090, 05089 SAO PAULO 10. Tel. 261.1920
Rua Senador Dantas 19 s/205-6, RIO DE JANEIRO GB.
Tel. 232.07.32

CANADA
Renouf Publishing Company Limited,
2182 St. Catherine Street West,
MONTRÉAL, Quebec H3H 1M7. Tel. (514)937.3519
522 West Hasting,
VANCOUVER, B.C. V6B 1L6. Tel. (604) 687.3320

DENMARK – DANEMARK
Munksgaard Export and Subscription Service
35, Nørre Søgade
DK 1370 KØBENHAVN K. Tel. +45.1.12.85.70

FINLAND – FINLANDE
Akateeminen Kirjakauppa
Keskuskatu 1, 00100 HELSINKI 10. Tel. 65.11.22

FRANCE
Bureau des Publications de l'OCDE,
2 rue André-Pascal, 75775 PARIS CEDEX 16. Tel. (1) 524.81.67
Principal correspondant :
13602 AIX-EN-PROVENCE : Librairie de l'Université.
Tel. 26.18.08

GERMANY – ALLEMAGNE
OECD Publications and Information Center
4 Simrockstrasse 5300 BONN Tel. (0228) 21.60.45

GREECE – GRÈCE
Librairie Kauffmann, 28 rue du Stade,
ATHÈNES 132. Tel. 322.21.60

HONG-KONG
Government Information Services,
Sales and Publications Office, Baskerville House, 2nd floor,
13 Duddell Street, Central. Tel. 5.214375

ICELAND – ISLANDE
Snaebjörn Jönsson and Co., h.f.,
Hafnarstraeti 4 and 9, P.O.B. 1131, REYKJAVIK.
Tel. 13133/14281/11936

INDIA – INDE
Oxford Book and Stationery Co. :
NEW DELHI, Scindia House. Tel. 45896
CALCUTTA, 17 Park Street. Tel. 240832

INDONESIA – INDONÉSIE
PDIN-LIPI, P.O. Box 3065/JKT., JAKARTA, Tel. 583467

IRELAND – IRLANDE
TDC Publishers – Library Suppliers
12 North Frederick Street, DUBLIN 1 Tel. 744835-749677

ITALY – ITALIE
Libreria Commissionaria Sansoni :
Via Lamarmora 45, 50121 FIRENZE. Tel. 579751
Via Bartolini 29, 20155 MILANO. Tel. 365083
Sub-depositari :
Editrice e Libreria Herder,
Piazza Montecitorio 120, 00 186 ROMA. Tel. 6794628
Libreria Hoepli, Via Hoepli 5, 20121 MILANO. Tel. 865446
Libreria Lattes, Via Garibaldi 3, 10122 TORINO. Tel. 519274
La diffusione delle edizioni OCSE è inoltre assicurata dalle migliori librerie nelle città più importanti.

JAPAN – JAPON
OECD Publications and Information Center,
Landic Akasaka Bldg., 2-3-4 Akasaka,
Minato-ku, TOKYO 107 Tel. 586.2016

KOREA – CORÉE
Pan Korea Book Corporation,
P.O. Box n° 101 Kwangwhamun, SÉOUL. Tel. 72.7369

LEBANON – LIBAN
Documenta Scientifica/Redico,
Edison Building, Bliss Street, P.O. Box 5641, BEIRUT.
Tel. 354429 – 344425

MALAYSIA – MALAISIE
and/et **SINGAPORE - SINGAPOUR**
University of Malaysia Co-operative Bookshop Ltd.
P.O. Box 1127, Jalan Pantai Baru
KUALA LUMPUR. Tel. 51425, 54058, 54361

THE NETHERLANDS – PAYS-BAS
Staatsuitgeverij
Verzendboekhandel Chr. Plantijnnstraat
S-GRAVENAGE. Tel. nr. 070.789911
Voor bestellingen: Tel. 070.789208

NEW ZEALAND – NOUVELLE-ZÉLANDE
Publications Section,
Government Printing Office,
WELLINGTON: Walter Street. Tel. 847.679
Mulgrave Street, Private Bag. Tel. 737.320
World Trade Building, Cubacade, Cuba Street. Tel. 849.572
AUCKLAND: Hannaford Burton Building,
Rutland Street, Private Bag. Tel. 32.919
CHRISTCHURCH: 159 Hereford Street, Private Bag. Tel. 797.142
HAMILTON: Alexandra Street, P.O. Box 857. Tel. 80.103
DUNEDIN: T & G Building, Princes Street, P.O. Box 1104.
Tel. 778.294

NORWAY – NORVÈGE
J.G. TANUM A/S Karl Johansgate 43
P.O. Box 1177 Sentrum OSLO 1. Tel. (02) 80.12.60

PAKISTAN
Mirza Book Agency, 65 Shahrah Quaid-E-Azam, LAHORE 3.
Tel. 66839

PHILIPPINES
National Book Store, Inc.
Library Services Division, P.O. Box 1934, MANILA.
Tel. Nos. 49.43.06 to 09, 40.53.45, 49.45.12

PORTUGAL
Livraria Portugal, Rua do Carmo 70-74,
1117 LISBOA CODEX. Tel. 360582/3

SPAIN – ESPAGNE
Mundi-Prensa Libros, S.A.
Castello 37, Apartado 1223, MADRID-1. Tel. 275.46.55
Libreria Bastinos, Pelayo 52, BARCELONA 1. Tel. 222.06.00

SWEDEN – SUÈDE
AB CE Fritzes Kungl Hovbokhandel,
Box 16 356, S 103 27 STH, Regeringsgatan 12,
DS STOCKHOLM. Tel. 08/23.89.00

SWITZERLAND – SUISSE
OECD Publications and Information Center
4 Simrockstrasse 5300 BONN. Tel. (0228) 21.60.45
Local Agents/Agents locaux
Librairie Payot, 6 rue Grenus, 1211 GENÈVE 11. Tel. 022.31.89.50
Freihofer A.G., Weinbergstr. 109, CH-8006 ZÜRICH.
Tel. 01.3634282

TAIWAN – FORMOSE
National Book Company,
84-5 Sing Sung South Rd, Sec. 3, TAIPEI 107. Tel. 321.0698

THAILAND – THAILANDE
Suksit Siam Co., Ltd., 1715 Rama IV Rd,
Samyan, BANGKOK 5. Tel. 2511630

UNITED KINGDOM – ROYAUME-UNI
H.M. Stationery Office, P.O.B. 569,
LONDON SE1 9NH. Tel. 01.928.6977, Ext. 410 or
49 High Holborn, LONDON WC1V 6 HB (personal callers)
Branches at: EDINBURGH, BIRMINGHAM, BRISTOL,
MANCHESTER, CARDIFF, BELFAST.

UNITED STATES OF AMERICA – ÉTATS-UNIS
OECD Publications and Information Center, Suite 1207,
1750 Pennsylvania Ave., N.W. WASHINGTON D.C.20006.
Tel. (202) 724.1857

VENEZUELA
Libreria del Este, Avda. F. Miranda 52, Edificio Galipan,
CARACAS 106. Tel. 32.23.01/33.26.04/33.24.73

YUGOSLAVIA – YOUGOSLAVIE
Jugoslovenska Knjiga, Terazije 27, P.O.B. 36, BEOGRAD.
Tel. 621.992

Les commandes provenant de pays où l'OCDE n'a pas encore désigné de dépositaire peuvent être adressées à :
OCDE, Bureau des Publications, 2, rue André-Pascal, 75775 PARIS CEDEX 16.

Orders and inquiries from countries where sales agents have not yet been appointed may be sent to:
OECD, Publications Office, 2 rue André-Pascal, 75775 PARIS CEDEX 16.

OECD PUBLICATIONS, 2, rue André-Pascal, 75775 PARIS CEDEX 16 - No. 42005 1981
PRINTED IN FRANCE
850/TH (77 81 05 1) ISBN 92-64-12247-8